The Triumph of Rationality

From Surgical Practice to Public Service

C. H. Leong

The Commercial Press

The Triumph of Rationality: From Surgical Practice to Public Service

Author:	Leong Che-hung
Cover photography:	Ducky Tse
Executive editor:	Chris Cheung
Cover design:	Cathy Chiu
Publisher:	The Commercial Press (H.K.) Ltd., 8/F, Eastern Central Plaza, 3 Yiu Hing Road, Shau Kei Wan, Hong Kong
Distributor:	The SUP Publishing Logistics (H.K.) Ltd., 3/F, C & C Building, 36 Ting Lai Road, Tai Po, New Territories, Hong Kong
Printer:	Elegance Printing and Book Binding Co. Ltd. Block A, 4th Floor, Hoi Bun Building 6 Wing Yip Street, Kwun Tong, Kowloon, Hong Kong

First Edition, First printing, July 2018

ISBN: 978 962 07 5797 6

Printed in Hong Kong

Contents

Preface

This is a story book; stories that are based on facts; stories about some of the changes that occurred in Hong Kong before and after the sovereignty transfer in 1997 that I myself had been involved in, and taken an active and leading role.

I write this book to remind myself of what a bumpy ride it had been and as an apology to my wife, my children and my friends for my neglect of them because of my devotion to public service.

Writing this book is easy – it only took me a few months of scrambling amid my many commitments. Getting it published is a challenge. I am grateful to my secretary Ms. Louisa Fu for deciphering my unreadable handwriting. I am grateful to Mr. Perry Lam for his excellent advice and succinct editing. I am also thankful to my former political assistant Ms. Kathy Chan for always being there to help when my memory failed me. Above all, I am in debt of the Commercial Press which gives me the greatest support by offering to publish the book.

C.H. Leong
2018

Prologue

The paternal influence

My father Dr. Leong Kam-leng (梁金齡) came to Hong Kong to study medicine from Singapore and graduated with a Bachelor of Medicine/Bachelor of Surgery (MBBS) from the University of Hong Kong in 1938. It was in the following year that I was born. Hong Kong, not the Lion City, thus became the place where my roots were planted.

University of Hong Kong

FACULTY OF MEDICINE

CERTIFICATE ISSUED FOR PURPOSES OF
MEDICAL REGISTRATION

This is to Certify that

LEONG KAM LENG

1. Passed the final examination for the M. B., B. S. Degree of this University in December, 1938.
2. That he is of good moral character
3. That this photograph is sure proof of his identity.

Dean

All the necessary fees have been paid.

Registrar.

Date 7th December 1938.

A proud graduate of HKU

Graduating medical students then were posted to different settings. As my father was not exactly a bright student, he was "exiled" to Aberdeen, then a poor fishing village where he was not just the only doctor, but probably the only person who could speak and understand English.

Then the Second World War broke out, and we left for Guangzhou to flee the Japanese occupation. I can still hazily remember that we led a reasonably well-off existence – my father was in charge of the hospital of Guangdong Province (廣東省立醫院). We travelled in private rickshaws most of the time. As far as I can remember, the hospital acquired an ambulance that for some reasons never worked. What I vividly remember is that almost every night we had to take shelter in the hospital avoiding air raids, and the devastation left behind by the bombs that fell elsewhere while the hospital was spared.

After the war ended, we returned to a dilapidated Hong Kong. By then there were seven of us taking up residence in a shared flat in Hak Po Street in Mong Kok in a room separating from our "neighbours" by panels that hardly extended to the roof.

But our lucky star was shining. My father was welcomed back to practise in Aberdeen. As the only doctor and the only person who spoke English, he also became the interpreter for Government officials and the local fisher folks.

He opened a clinic in Wu Nam Street. Medical practice then was spartan to say the least. My parents had to carry the daily provisions of medicine in rattan baskets everyday as they travelled

Serving a bridge between the Government and local people

to Aberdeen by bus remodelled from trucks which, at that time, was the only means of transport to and from Aberdeen.

His practice was comprehensive – from treating cough and cold, to infected fingers from fish hook puncture, to delivering babies.

He was thus a doctor-at-large. Everyone in Aberdeen knew him, for whenever they were sick, there was no one else to turn to as Aberdeen then was an isolated part of Hong Kong.

What sort of doctor my father was came through most clearly in an incident that took place under tempestuous weather. Typhoon signal no. 8 was hoisted and there was a fisherwoman in labour on a junk at the Aberdeen harbour. Without thinking twice, my father and my mother, a nurse, went out on a sampan and in the middle of a rough sea, delivered a five pound baby, a first born, to the delight of his grandparents and parents. A son, a first born, and therefore a jewel in the eye. And the reward – the fresh catch of the day, a catty of shrimps and subsequently a couple of specially prepared salted fish.

There was nothing my father took more seriously than his duties as a doctor. It was the first day of the Lunar New Year and he was called to certify an elderly fisherman who died on a fishing junk. Without hesitation and undeterred by the superstition that associated it with bad luck, he jumped onto a sampan and performed the necessary ritual. He never asked for a fee, but was rewarded with a red packet of 10 dollars.

My father was mesmerized by the community and, in return, he was considered an "honourary Tanka person" (蜑家佬).

Together with my sisters and brothers, I spent most of the weekends when we didn't have to go to school in his clinic in Wu Nam Street, which was about the only "quality time" we had to be with our parents. Not much parental care was given to us for both my parents had to work hard to deal with their heavy workload. Yet they were most caring and they taught us by example. In this aspect, we were no different from many other children then. For most big families living from hand to mouth, the parents had to work very hard to provide for their children. Those were the days of simplicity when people still believed in "the Lion Rock spirit" and the virtues of hard work.

My father subsequently opened another clinic at 74 Queen's Road in Central. It was under such circumstance that I gradually realized what taking up the medical profession is all about –

A family portrait. My father is fourth from the right

My father's clinic at the junction of Pottinger Street and 74 Queen's Road Central (See signboard in the top-left corner) (Courtesy of Mr. Luk Hon-yat)

commitment, caring, patients first and their recovery taking precedence over everything else. My parents were therefore my role models and they taught and inspired me to take up medicine as my career.

Chapter 1

Against all odds

Surgery is a challenging specialty. It is also a demanding specialty. Physically-demanding, as the surgical procedures could be tedious. Mentally-demanding, for oftentimes the surgeon has to make quick and consequential decisions. If you do nothing, the patient will die; if you do something, he might have a chance. Yes, it is a matter of life and death. You weigh all the pros and cons, then you make a decision and you never regret. It toughens you and your character. You "get on with it" and you don't "dilly-dally".

Life as a trainee surgeon then was by no means easy. There were no "standard working hours" and nobody talked about "work-life balance". You were off duty only when all the work had been done, with no patients' complaints to deal with. As one of the two doctors in the entire surgical department treating 200 patients (the other being Dr. A.E.J. van Langenberg) , I barely had time to leave for a haircut.

Ward round started at 8am, and was over by 10am. Thereafter the professor usually left for boating pleasure at the then Royal Hong Kong Yacht Club, and senior doctors took off for surgery in private hospitals. The two of us were left to take charge of all the patients and emergency admissions.

On the days designated for operations, one of us had to assist the professor and senior surgeons to perform surgeries. The other one had to do all the preparations for the patients scheduled for surgery: setting up blood lines, checking their status, etc.

Yet I never uttered a single word of complaint, nor did anybody else. For me, surgery is a calling, not a job. Surgery may be a science, but its practice is an art: an art in making decisions, an art in communicating with patients; an art in instilling confidence in the patients and their families; and finally an art in following surgical procedures.

Since it is an art, exposure and experience matter – the more you see, the more you learn, and "practice makes perfect". As my former chief of surgery once said to his blue-eyed trainee who boasted he could remove a patient's stomach in 45 minutes, "If I train a monkey long enough, it will be just as good."

Indeed, a good surgeon is not only a doer. He must also be a thinker and an innovator. Standard surgical techniques are well tried and time-honoured, and can be relied upon to give evidence-based good results. Yet treatments must be adapted to the passage of time, the needs of the patients and cultural differences. Surgeons must keep an open mind about innovations that put patients on the fast track to normal lives.

Treatment of cancer of the bladder, presenting with blood in the urine, is a case in point. In days gone by, the symptoms were usually interpreted as little more than a simple infection of the urinary tract that called for prescription of antibiotic. The seriousness of the condition was thus overlooked. The recurrence of these symptoms, however, meant the cancer had reached an advanced stage and the only solution was to remove the bladder altogether.

Unfortunately, that would also mean that the patient would be deprived of a receptacle to hold his urine which could only be drained through a hole in the lower part of his body. With his clothing occasionally soiled by urine, the patient gave off a weird odour. This, in the 60's and 70's was something to be frowned upon. The social stigmatization was so strong that the patient would rather suffer the consequence of an advancing bladder cancer and ultimately succumb, than the humiliation of being a social outcast.

Can we reconstruct a bladder using another hollow organ? A segment of large bowel had been tried to less than desirable effect – the contraction was poor, producing a weak urinary stream; the large bowel produced a lot of mucus, often clotting the urinary flow; the large bowel absorbed acid from the urine producing acidosis in the long term. What about a segment of the stomach?

The idea was thus conceived!

Was it technical feasible? The stomach has a thick muscle wall so it should also have a strong contraction. But which part of the stomach could be used? The upper part (body) produced a lot of acid and it may cause irritation or even excoriation of the urethra. The lower part (antrum) might work.

While there were then no ethics committees in hospitals in the late 60's and early 70's, it would still be inconceivable to use patients as "guinea pigs". Evidence base must be established – and I resorted to do a trial using dogs.

Dog lovers and animal rights advocates can rest easy. Every dog used in the trial was properly anaesthetized and had received proper surgical procedures. The laboratory I used, however, was spartan. There was only one technician in the daytime. Oftentimes, I had to return to the dog laboratory at night after social functions I had to attend, often in formal attire, to give post-operative care to my "precious" dogs – clysis and fluid replacement; introducing catheters to drain the "reconstructed" bladder, etc.

All the hard work paid off as the experiment proved to be a success. The "stomach" could be used and it had many advantages over other hollow organs of the body. Not only was it technically feasible, but it also had strong muscle contraction and produced a good urine flow. There was no adverse effect on the dog receiving the operation, and its mildly acidic urine effectively counteracted bacterial growth.

I went on to perform the procedure on patients whose bladders had been removed entirely for the treatment of cancer.

It was a global first

It was such a success that the procedure was subsequently used to enlarge bladders which had contracted to the size of a "thimble" due to tuberculosis of the urinary system – a common condition in the 60's and 70's. The procedure is now commonly known as "augmentation gastro cystoplasty".

Today, while tuberculosis of the bladder is rare, long-term ketamine addiction produces the same complications and would benefit from augmentation gastro cystoplasty.

Academic surgery is an honourable act. You develop a new and better surgical procedure but will never try to get a patent. On the contrary, you want as many people as possible to benefit from your innovation. If your work has true scientific value, you can write it up and the paper will be admitted for publication in important medical journals and you will get honourable mention by prestigious medical organizations.

The work on gastrocystoplasty was considered a breakthrough both in technology and in scientific research, and I was named a Hunterian Professor of the Royal College of Surgeons of England. What it entailed was that I had to fly to London to deliver a professorial lecture.

It was a great honour and an eye-opener. On the big day, I was led into the Edward Lumley Hall of the Royal College of Surgeons of England in Lincoln's Inn Field in Holborn, London, by the three "wise men" in full academic regalia – the President of the College, the Senior Vice President and the Registrar. Later, they sat straight-faced in the front row while I delivered the lecture. When the event was over, the trio led me into the President's Chamber where I was offered a congratulatory sherry and a cheque in the amount of one guinea.

Nevertheless, I take pride in being the third person from Hong Kong, the first Hong Kong-born Chinese and the first Hong Kong University graduate to be so honoured.

Ultra-major surgery may lead to transient kidney failure. Before the kidneys start working again, the doctor has to keep the patient alive by artificially removing the "toxic waste products" of metabolism using an artificial kidney – a process called dialysis. In 1963, with a donation from the Rotary Club, an artificial kidney was installed in the surgical department of Queen Mary Hospital, and I became the first person to establish and take charge of a dialysis unit in Hong Kong.

The unit subsequently expanded to provide total replacement therapy – chronic dialysis for chronic renal failures, i.e. patients without functioning kidneys, with Professor Richard Yu (余宇康).

The role of doctors in society is to save lives by providing the best possible medical treatment, yet the role I had to play was more like God when I was given the charge of the dialysis unit. Once a patient had terminal kidney failure and was accepted for dialysis, he would need permanent dialysis from a machine 2-3 times a week. Such intensive use of the machine meant that it could only benefit a limited number of patients.

The question was who should get the benefit. It was a tough call to make. Imagine a situation where there were two patients suffering from kidney failure and you could only admit one of them. To admit a patient was to give him a "kiss of life", and to

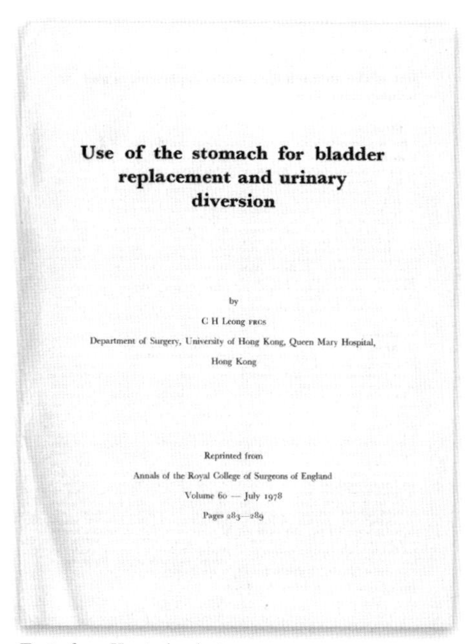

Use of the stomach for bladder replacement and urinary diversion

by

C H Leong FRCS

Department of Surgery, University of Hong Kong, Queen Mary Hospital, Hong Kong

Reprinted from

Annals of the Royal College of Surgeons of England

Volume 60 — July 1978

Pages 283—289

Text of my Hunterian lecture

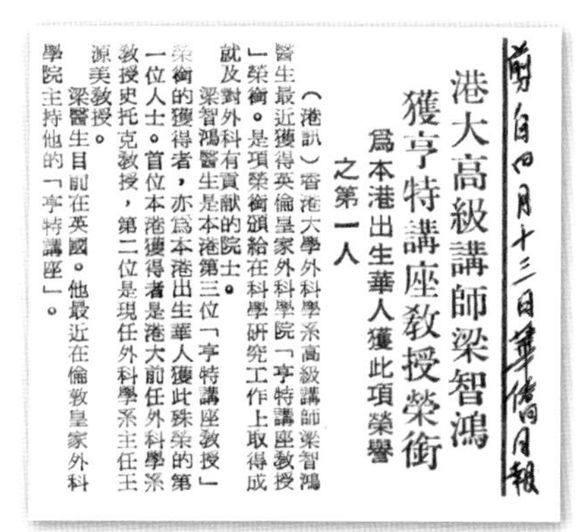

港大高級講師梁智鴻獲亨特講座教授榮銜

爲本港出生華人獲此項榮譽之第一人

（港訊）香港大學外科學系高級講師梁智鴻醫生最近獲得英倫皇家外科學院「亨特講座教授」榮銜。是項榮銜頒給在科學研究工作上取得成就及對外科有貢獻的院士。

梁智鴻醫生是本港第三位「亨特講座教授」榮銜的獲得者，亦爲本港出生華人獲此殊榮的第一位人士。首位本港獲得者是港大前任外科學系教授史托克教授，第二位是現任外科學系主任王源美教授。

梁醫生目前在英國。他最近在倫敦皇家外科學院主持他的「亨特講座」。

Hunterian honour for HKU surgeon

SENIOR Lecturer in the Department of Surgery, University of Hongkong, Dr C.H. Leong, was recently named Hunterian Professor of the Royal College of Surgeons of England. The Hunterian Professorship is awarded by the College to fellows or members who have done work of scientific value and contributed to the progress of surgery.

Dr Leong is the third Hongkong recipient of this award, and the first local-born Chinese to be so honoured. The first two recipients were also from the University of Hongkong: Professor Francis Stock, former Professor of Surgery; and Professor G.B. Ong, the present Professor of Surgery.

Dr Leong is now in Britain, where he recently delivered his Hunterian Lecture entitled 'The use of the stomach for urinary diversion and bladder replacement' at the Royal College of Surgeons in London.

News clippings kept by my father

In full academic regalia at the Royal College of Surgeons

reject a patient (because there was no available facility) was to "sign his death warrant". You became literally his "executioner". Imagine again: patient A was a 35-year-old bread-earner, and his wife gave birth to a son three months ago. Patient B is a 50-year-old singleton with no family to support. Both were economically active persons contributing to the society. Yet the dialysis unit could only admit one more patient. Which of the two patients would you admit?

You can of course ask for more machines, but the bottleneck

With HKU classmates

will always be there. There will always be more patients than machines. All you can do is to bite the bullet and do your best to make a conscientious, sensible and fair decision. It may turn out to be not exactly the best decision and you cannot please everybody, but your conscience is clear. You can still sleep at night without having bad dreams.

The science and practice of surgery never comes to a standstill. The surgeon's quest for the continual better treatment therefore never ends.

The news of the success of renal transplant in the 1954 opened a totally new horizon for us. From then on, we can treat more patients with non-functioning kidneys, give them a new and functioning kidney and return them to a much better life. Gone are the days of patients submitting themselves 2-3 times a week for dialysis (4-5 hours each time). They can now literally return to their daily routines and travel without being "attached" to a machine.

The procedure was new to us, so we went back to the animal laboratory to perfect our technique in dogs.

Back in 1969, if you were told that you would be the first patient to receive and benefit from a kidney transplant you might greet the news with more doubts and questions than joy and relief. Am I just a guinea pig? Will the surgery succeed? And will the new kidney function? What will my life be after the transplant? Worse still, there was no way that you could be sure when the procedure would take place – all at the mercy of a possible donor. You would be stuck to your landline (there was no mobile phone then), you had to return to the hospital immediately once a possible potential donor was identified, only to be repeatedly disappointed as his or her family refused to give their consent.

Nor would it be an easy decision for the donor's family. Put yourself in their shoes. Your daughter left home in the morning happily, saying that she would return for dinner. At noon, the police informed you that she had a traffic accident and was now in hospital in critical condition. You rushed to the hospital, only to be told by the doctors that there was no hope. You begged, you pleaded and prayed that the doctors would do something to save her life. In a state of total despair, you were approached by another team of doctors who asked for your consent to donate your daughter's kidney when she passed away to save somebody's life.

A first-generation dialysis machine (Kolff twin coil)

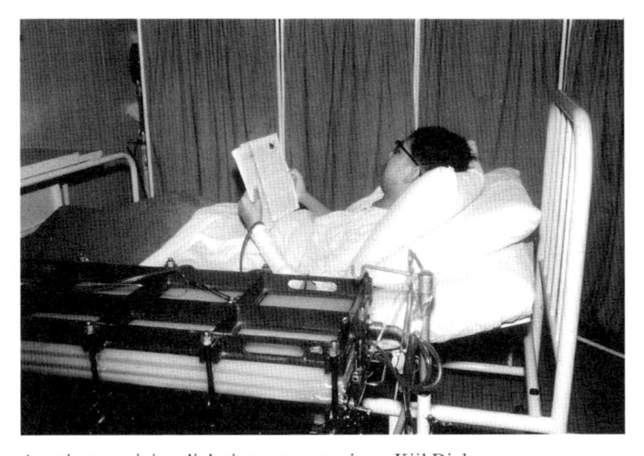

A patient receiving dialysis treatment using a Kiil Dialyser

"It is my daughter's life that I am concerned with, not somebody else's." You lost control and slapped the doctor. What went through your mind then, though totally groundless but entirely understandable, was that the doctors were waiting for your beloved daughter to die so that they could snatch her kidneys like "vultures". That's why they wouldn't do their best to save your daughter's life.

The doctors themselves similarly had an impossible task. It is a common misunderstanding that doctors want the glory of doing a transplant. The truth is their ordeal begins when consent is given by the patient's next of kin. Back in the 60's, brain death was *not* normally accepted by society as death. Doctors therefore had to wait for the patient's cardiac death (when the heart stops) before they could remove his kidneys. As soon as the heart stops, however, there will be no blood supply to the different organs and in a short time, the organs would be damaged. The doctors therefore had to stay vigilant and close to the potential donor all the time. As soon as the patient's heart stopped, they had to immediately perform cardio pulmonary resuscitation manually, then take the donor to the operation room while continuing with the resuscitative process until the donated organs were removed and properly treated. And all this had to be done in the presence and under the watchful eyes of the patient's relatives. It was a heart-breaking, physically-demanding and mentally-draining process.

Yet no doctors can't help but feel jubilation, rejoice, and a profound sense of satisfaction when they join the blood vessels to the recipient and released the clamps. The donor kidney turns pink. Everyone in the operation room – the surgeon, the assistant, the anesthetist, the scrub nurse, the circulating nurse all held their breath. Five minutes later, peristalsis began to be seen in the donor's ureter, and the first spurt of urine appears – representing the beginning of a new life!

On January 8, 1969, history was made. The first kidney transplant was successfully done in the University of Hong Kong's Department of Surgery at Queen Mary Hospital. I was a member of the team.

We were asked to keep a low profile. "Do not publicize the event," said Professor G.B. Ong (王源美), the Head of Surgery. We obliged and the transplant team kept quiet while he, as expected as the head of the department, held a press conference the first thing in the morning. The following was an excerpt from the press release issued by the hospital:

> *A kidney transplant operation was successfully performed in Queen Mary Hospital last night on Mr. Ng Ho Bun who has been in hospital since September 1968.*
>
> *The kidney was taken from a 19-year-old girl who died yesterday with her parents' consent which was given in writing.*

It was the sweet taste of success and, naturally, spurred the enthusiasm for more transplants.

Thus began the "long fight" for organ donation. To this end,

I founded both the Hong Kong Society of Nephrologist and the Hong Kong Kidney Foundation.

The medical profession was burning with the enthusiasm to do more kidney transplantations – after all, dialysis sustained life, but to receive a "new" kidney means a new life – a life that the recipient can and will return to almost total normality. This enthusiasm, unfortunately, was not shared by that of the public.

The public's reluctance to donate organs at death is not hard to understand:

- There is a superstitious belief, widely held by local people, that if a person is *not* buried with his body intact, he will be condemned to the seventh level of hell;
- Many also share the view that if I were to give my consent to donate my organs after death, and if I meet with an accident, the doctors will not do their best to save me.

Such misconceptions can only be countered by education. We mounted a territory-wide promotion campaign which encouraged citizens to sign a "kidney donation card" and carry it in their wallets. The response, however, was poor to say the least.

Experience from other parts of the world – notably Singapore, Spain and Northern Europe – have shown that the use of an "opt out" system leads to a significant increase in cadaveric organ donations. Under the system, everyone is presumed to be willing to donate their organs upon death unless they signed something to the contrary. It's called the "opt out" system because you can "opt out" any time during your life time.

I introduced a debate at the Legislative Council to publicize the system and to gauge public opinion. To my surprise, it was met with extensive opposition on the ground that "opting out" was against "human rights".

This is beyond me. When a person can always say no, how can his rights be violated?

In Singapore, the implementation of the "opt out" system had led to an increase in people giving consent to organ donation – citizens feel that since they can always choose to "opt out", why not "opt in"?

Since then, medical developments had seen success in other organ donations and transplantations, such as liver and heart. As a result, many organ donation campaigns in Hong Kong had been launched by the Government and non-governmental organizations – the Rotary Club, medical bodies, the Hong Kong Kidney Foundation, etc. Regrettably, the response of the public had been less than enthusiastic for, I believe, two reasons:

- the campaigns lack sustainability;
- living donors are shown to have rather low morbidity and mortality, and family members of patients and good Samaritans are willing to help, especially in liver transplantation.

Of late, the idea of "opting out" is being discussed again. It remains to be seen how society, now even more politicized than before, will respond. The stumbling block is of course the society is now over-politicized.

Another impediment to organ donation is of course the law. A person may have signed all the necessary documents to donate his organs after death. Yet, legally, once a person dies, his body belongs to his next of kin who could act against his expressed will to donate his organs. It is therefore imperative that the would-be donor should make absolutely clear to his next of kin his wish which, hopefully, will be honoured when the heartbreaking time comes.

Chapter 2

There is no place like home

In an attempt to improve the results of transplantation by learning more about tissue rejection, and to develop urology as a separate specialty in Hong Kong, I accepted a China Medical Board Scholarship to work for a year at the University of California, Los Angeles (UCLA) on the west coast of USA – under Professor Paul Terasaki (寺崎一郎) – a renowned transplant scientist who discovered human leukocyte antigen (HLA), and Joseph J. Kauffman, a master urologist, humourist and gifted pianist.

It was an eye-opening experience. Everyone there was driven and industrious. Ward round started at 5:30am, which was usually followed by a hearty breakfast – six layered pancakes, four fried eggs with ham, and lots of "French fries". Then it's time to "scrub up" and get ready for the "OR" (operating room).

Trainees or fellows from Hong Kong were always welcomed. They worked hard, seldom complained, and did all the daily chores – catheterizing bladder, etc. They were knowledgeable too, with usually 4-5 years of experience in Hong Kong doing different sorts of duty in the hospital.

Life on the fast lane in Los Angeles

The United States was an eye-opener in another aspect – as a land of plenty. At 5am every morning the six-lane highways on both sides of the road would be lined with monstrous motorcars each with engines of six-litre capacity, usually carrying only the drivers with no passengers. Yet there were millions who did not even receive proper medical care when they got sick.

I was once on an ambulance on a kidney harvesting mission for transplantation, when we picked up a man who apparently had had a heart attack. The ambulance frantically scouted for hospitals in the vicinity to send the patient to. Then the following conversation took place:

Does he have insurance?

No.

Don't send him here.

And the line went dead.

No doubt the United States is a land of opportunities. I received attractive offers to stay and work in the US after my sabbatical leave was over. Joseph J. Kauffman asked me to join UCLA as the Head of Experimental Surgery; Rush Medical Centre in Chicago offered me a job as transplant surgeon, and Ian Thompson, a professor in urology from Mid-West (Columbia) offered me a "second in command" position in the medical school of a new university in Florida. All these offers gave me the chance to become a US citizen. I turned down the laboratory job as I believed my destiny was to become a surgeon for people. Experimental surgery is important as many medical breakthroughs start off with research, my passion is to treat people and that has to be my career.

I turned down the lucrative job from Chicago as I could not stand its strong wind and the cold weather. Perhaps I went to the job interview at the wrong time of the year – end of November. The job in Florida did not materialize, for sadly Dr. Thompson developed a heart attack from which he never fully recovered. What really made me say no to these offers, however, was that the strong feelings I have for my homeland and home city. Yes, I could get rich and perhaps even famous on foreign soil, but my roots were in Hong Kong and my passion was to treat Hong Kong people and my compatriots. I had come to the United States to learn and improve myself as a surgeon, so as to do a better job in treating my patients in Hong Kong and China. Besides, I can't stand being treated as a second-class citizen on foreign soil.

Of the so-called "Asia's Four Little Dragons", Singapore, South Korea and Taiwan may have surpassed Hong Kong in certain areas of economic development. Yet in medical practices and development, at least, Hong Kong remains the leader and for good reasons. Hong Kong has two medical schools. The medical school of the University of Hong Kong has a long history; we have adopted the time-honoured British system to train all-round doctors and the medical profession enjoys a high degree of professional autonomy. Our scientific research and academic standing are recognised to be in par with if not exceeding other international medical schools. In fact, many medical doctors in Singapore and Malaysia were our graduates.

The chance for me to use my expertise to serve my compatriot outside Hong Kong came in the late 1970's. At that time, while Hong Kong boasted a good health care system, Macao, a small Portuguese enclave some 40 miles away, lagged very much behind. The majority of its residents including high ranking government officials came to Hong Kong to seek medical care for anything more serious than a common cold.

I was asked by the directors of the Kiang Wu Hospital (鏡湖醫院) to assist in setting up a surgical department. So every Saturday after finishing work in my clinic in Hong Kong, I jumped on to a hydrofoil to travel 1.5 hours to Macao where I would be picked up at the pier and taken to the Hospital to perform surgery, give lectures and provide advice on setting up a surgical department and improving the operation room facilities. The Hospital Board

At Kiang Wu Hospital (鏡湖醫院) with Dr. Chan Yu-kai (陳汝啟)

under Dr. Ko Lin (柯麟) and Henry Fok (霍英東) gave me a free hand, and Director Dr. Leung Chi-fai (梁志輝) and Dr. Chan Yu-kai (陳 汝 啟) who acted as my second-in-command, were very helpful. If the work finished early, I would return on the same night to Hong Kong on Saturday and if not, on Sunday afternoon. I did this every weekend and subsequently once a month from 1978 to 1987. It was a little tiring but also gratifying. A surgical department, though rudimentary, was established to serve the people of Macao. Much more work needs to be done to raise its standards. Macao may now boast a gambling and entertainment industry that rivals Las Vegas, but its health care system still lags far behind.

Since 1949 when the People's Republic of China was

established (The Bamboo Curtain) , the door of the nation had been closed to foreign influence and activities. The Cultural Revolution and the "Gang of Four" did not make things any better. As a result, in the field of medical practice, China in 1979 was 40 years behind the more medically-advanced parts of the world, such as Hong Kong.

I was asked by the Sun Yat-sen University in 1979 to travel to Guangzhou to perform a transurethral resection of the prostate (TURP) , a procedure by which the prostate was removed via the urethra) . Bringing along a new resectoscope as a gift, I travelled on a 15-minute flight to Baiyun International Airport. At the hospital, I had to tune the Bovie cautery diathermy machine used to perform the procedure myself and performed the operation under the watchful eyes of many in a crowded, natural air-ventilated operation room. Nevertheless, the surgery was a success.

In the evening after a "feast", I relaxed with a glass of Maotai and spoke to the chief surgeon. I told him I was amazed that TURP, which had been performed in Hong Kong and the rest of the world for many years, was not done in China.

It was then revealed to me that an American surgeon did the operation in Guangzhou a few years ago but the patient succumbed afterwards for some other reasons. The operation was since banned by the administration.

It was a shock to me that professional autonomy, sacrosanct in Hong Kong, could be overridden by politics and administrative expediency like this in China. In the days to come before the transfer of sovereignty of Hong Kong, I took it upon myself to protect the autonomy of the medical profession after 1997. This became my self-imposed mission.

Not much later I was invited to Quanzhou (泉州). Despite the language barrier, I again accepted the invitation. This time, I travelled by boat to Xiamen and then four hours by car to Quanzhou. As I arrived, I was given a standing ovation by some 200-300 people headed by the party secretary and mayor. Then my heart sank as I was led to examine some 40 patients with enlarged prostate whose urine was drained by makeshift catheters. The conditions of many were complicated by having bladder stones. The makeshift catheters were red rubber "nelaton" catheters tied to a condom that acted as a retention balloon. It was a horrible, heartbreaking sight. While we in Hong Kong were enjoying all the comfort and facilities of modern medicine and surgery, our brothers and sisters just north of the border were subjected to "stone-age" treatment for their sufferings. It was at that moment that I made up my mind to do my very best to help my fellow Chinese in the Mainland. This had since become my quest.

Other invitations brought me to Beijing and Shanghai and I took these opportunities to learn more about the New China. I also renewed my acquaintances with the late Wu Jieping (吳階平), the father of Urology in China, and his nephew Wu Decheng (吳德誠), a fellow urologist at the then Capital Hospital (首都

醫院）. We did renal transplantations and lithotripsies together. I remember one lecture trip to Urumqi vividly as it demonstrated the trials and tribulations of travelling in China. It was in January, 1981 and I took a flight from Guangzhou to Urumqi. After a delay of nearly six hours, we were up in the air when the pilot announced that it was too late and too dark to continue flying and we had to land in Lanzhou（蘭州）. The airport in Lanzhou then in 1981 was no more than a makeshift, spartan big hut and it was over 70 kilometres from town. We were forced to stay at the air terminal. The flight to Urumqi was subsequently cancelled for four days as the city was covered by heavy snow.

For four days I was marooned at the air terminal, with no shower, and no change of clothing. What did I have for food? The majority of the travellers were Uygur and Muslims who only take "blessed" food. As a result, hard-boiled eggs alone were served with each meal. I lost count of how many of them I had taken, but the sight of hard-boiled eggs, today, now some 30 years later, still sends shivers down my spine.

Every time I took these trips, I always brought with me a new resectroscope and 50 pieces of FG 20, three-way foley catheters as my gifts to my hosts.

Recently, I took another trip to Lanzhou on an AIDS mission and was overjoyed to discover how modern and impressive the airport has now become. That also means I am unable to relive my experience.

This testifies to the incredible progress China has made since

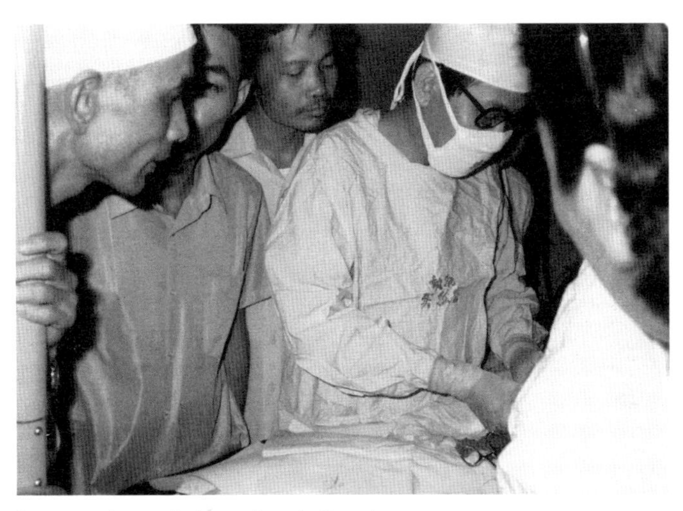

Demonstrating surgical procedures in Quanzhou

With Dr. Wu Decheng (吳德誠) at the Capital Hospital (now Beijing Union Hospital)

At the Lanzhou airport

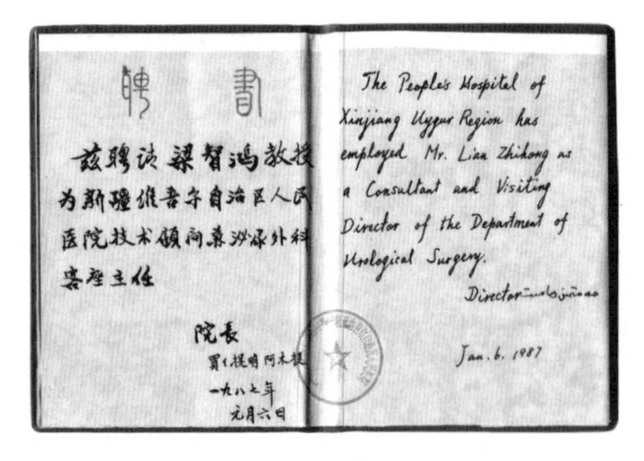

聘書

兹聘请梁智鴻教授
为新疆维吾尔自治区人民
医院技术顾问兼泌尿外科
客座主任

院長
買々提明阿木提
一九八七年
元月六日

The People's Hospital of
Xinjiang Uygur Region has
employed Mr. Lian Zhihong as
a Consultant and Visiting
Director of the Department of
Urological Surgery.

Director

Jan. 6. 1987

Letter of appointment issued by the Urumqi government

the fall of "the Gang of Four" and the opening up of the country under the leadership of Deng Xiaoping. To cite but one more example, China now has one of the most advanced highway and high-speed railway network systems in the world. The sleeping giant has awakened.

Chapter 3

Preparing the professionals in Hong Kong for the transfer of sovereignty

As 1997 drew near, Hong Kong people began to get jittery. What will Hong Kong be like under communist rule? Will Hong Kong be turned from a free society into one ruled by an autocratic regime? Will the travel freedom of Hong Kong citizens be curtailed? The medical profession was similarly worried. Many rushed to acquire citizenship of other countries. To those who could fulfill the requirements for citizenship, Canada, United States and Australia became their safe havens. The so-called "astronaut phenomenon" was getting increasingly common – while their families stayed in their safe haven countries to comply with the immigration requirements, the husbands as breadwinners stayed in Hong Kong to work. Some of our colleagues were said to have bought 52 round-trip air tickets to Australia at one go (at discount rate for bulk purchase). As a routine, they would take off from Hong Kong on Friday night, return on Sunday evening and resume work on Monday. Chartered flights to Sydney, Melbourne and Canberra were arranged to tie in with the registration dates for medical practices in those cities. Long queues appeared outside the Singapore Embassy in Hong Kong, when it was announced that citizenship would be granted to investors. Many "buy" passports to countries like Central America as a last resort to leave Hong Kong.

Disappointingly, Britain refused to honour her commitment to Hong Kong people who had legitimate claims to be her subjects by birth. This contrasts sharply with the situation in Macao, an enclave 40 miles from Hong Kong under Portuguese rule. Like Hong Kong, Macao was to be returned to Chinese sovereignty. But unlike the British Government, the Portuguese Government offered, seemingly without hesitation, Portuguese citizenship to the people of Macao. Hong Kong people were given a BNO (British National (Overseas)) or BDTC (British Dependent Territories Citizens) passports. These were simple travel documents with no right of abode in Britain or anywhere.

On the eve of the transfer of sovereignty to China, many Hong Kong people felt totally sold out by the British Government and who could blame them?

When the British Foreign Secretary Geoffrey Howe came to Hong Kong to "sell" the Joint Declaration to the Hong Kong people, a few of us set up a group called "ROAD" (Right of Abode Delegation) . The goal was not to help secure the right of abode for Hong Kong people – that would be a lost cause – but to highlight how unjustly we were being treated. Many of the members in the group are still working to make Hong Kong a better place, including Albert Cheng (鄭經翰) , Robert Tang (鄧國楨) , Donald Yap (葉天養) , Francis Yuen (袁天凡) and myself.

As a member of the Legislative Council, I commuted a few

times to Whitehall on my own, to lobby the British Government to give Hong Kong people what they deserved. It didn't take me long to realize that it was not the Hong Kong British subjects that the Foreign and Commonwealth Office was concerned with. They were worried that should the flood gate be opened, millions of former British subjects from Africa would swamp their country.

History of course had shown that all these worries were groundless. China proved to be steadfast in her commitment to the "One country, Two systems". The Hong Kong SAR passport subsequently issued by the Hong Kong SAR Government guarantees Hong Kong people's right of abode in Hong Kong and serves as a more superior travel document with visa-free access to, at present, 162 countries and territories.

My concern lied elsewhere. It has always been my conviction that the autonomy of the medical profession must be jealously guarded and the right to determine the standard of the services it provides to citizens belongs to the profession and the profession only. Don't get me wrong. Professional autonomy is not professional protectionism, but is the one thing that all professions cannot do without in their quest to provide the best possible services to the public. Only doctors themselves can determine whether the service of a doctor to his patient is up to standard. Only doctors themselves can determine whether a treatment is acceptable. Only doctors themselves can determine whether five years of medical training are adequate to turn a medical student into a doctor. Similarly, only airline pilots themselves can determine whether a new man should be given a license to fly a plane. Airlines pilots cannot tell doctors how to treat a patient, just as a doctor cannot tell a pilot how to fly a plane. In the same way, professional standards cannot be determined by the government.

In Hong Kong before 1997, the standards of doctors were determined by the Medical Council of Hong Kong and was chaired by the Director of Medical Health Services – a civil servant. The Council's membership included representatives among others from the Britain Medical Association (Hong Kong Branches) and medical members of Her Majesty's Armed Forces.

Two areas are distinctly unacceptable. Firstly, to have a medical council that determines professional standards chaired by a government official violates the spirit of professional autonomy, for professional standard could be compromised by administrative expediency.

Secondly, after the change of sovereignty, the Britain Medical Association and Her Majesty's Armed Forces would have *no* role to play in Hong Kong. The composition of the Medical Council of Hong Kong and the Medical Registration Ordinance that determines the Council therefore must be amended without delay.

Upgrading the Medical Council – amendment to the Medical Registration Ordinance

On assuming the role of an elected member of the Legislative Council, I moved to amend the Medical Registration Ordinance to replace the Director of Medical and Health Services as the

Council's *de facto* chairman with an elected member of the Council. This was to enhance the autonomy of the medical council. With the support of then Governor David Wilson (Now Lord Wilson) who was also President of the Legislative Council, it was done without much difficulties.

But to remove the Britain Medical Association from the membership of the Medical Council was an uphill battle. To pull it off, I fought for the Presidency of the Britain Medical Association (Hong Kong Branch). It was a long shot. I was a newcomer competing with a well-respected, much liked stalwart, Dr. Henry Lee (李福權). It was like David against Goliath but I won. I became the President of both Hong Kong Medical Association and the Britain Medical Association (Hong Kong Branch). That put me in a favourable position to convince the local members of the Britain Medical Association and its parent organization in the United Kingdom to give up their seats in the Medical Council. Similarly, the representative of Her Majesty's Armed Forces representative was removed from the Council's membership. All these changes formed part and partial of the amendments to the Medical Registration Ordinance.

A major battle was won, and the Medical Council became a totally professional autonomous body composed of locally-registered medical practitioners. There were lay members to monitor the medical profession's practice on behalf of the public. This is a major role a professional council should play. I will return to this issue in chapter 6.

Professional autonomy

Professional autonomy covers much more than just standards of practice. It safeguards the rights of Hong Kong professionals to participate on their own in international professional bodies and the rights of Hong Kong Professional Associations to join international bodies under the name of Hong Kong. This applied to all professions then. Hong Kong has always participated in international professional meetings on her own, independent of her sovereignty country. In the eyes of the international profession bodies, we are always capable of representing ourselves, independently and on our own.

We have earned the respect of the international professional bodies with our high professional standards, long association with international bodies, and proficiency in the English language.

This hard-earned independence would be vital if Hong Kong was to continue and develop its international networks and relationships after 1997. It was important to China as well for our global connections would help China's professional bodies gain international acceptance and recognition.

One example was the acceptance of the Chinese Medical Association (CMA, the official medical organization in China) by the World Medical Assembly (WMA) as a member.

In 1982, Hong Kong played host to the annual conference of WMA. As the President of the Hong Kong Medical Association, I took the opportunity to introduce CMA under the Presidency of

Wu Jieping to be a member of the Assembly.

So vital was professional autonomy to the development of Hong Kong and its professions that I was convinced that its principles must be enshrined in the "Basic Law".

The task fell on the shoulders of the various professionals sitting on the Basic Law Consultative Committee, among them: Donald Yap (葉天養) (lawyer), Dennis Chang (張健利) (barrister), Jeffrey Tsang (曾翼生) (dentist), Edward Ho (何承天) (architect), Michael Mann (surveyor), Raymond Ho (何鍾泰) (engineer), Bosco Fung (馮志強) (planner), Peter Wong (黃匡源) (accountant) and myself (medical). Led by the late Dr. Raymond Wu (鄔維庸), a member of the Basic Law Committee and a close friend of Beijing, we made numerous representations to Beijing, meeting in particular Ji Pangfei (姬鵬飛) and Li Hou (李后), stressing to them the importance of "professional autonomy". Above all, we insisted that the principle of professional autonomy is definitely in line with the principle of "One Country, Two systems".

Our message must have struck a chord. Ultimately, the following clauses formed part of the Basic Law:

Article 142

The Government of the Hong Kong Special Administrative Region shall, on the basis of maintaining the previous systems concerning the professions, formulate provisions on its own for assessing the qualifications for practice in the various professions.

Persons with professional qualifications or qualifications for professional practice obtained prior to the establishment of the Hong Kong Special Administrative Region may retain their previous qualifications in accordance with the relevant regulations and codes of practice. The Government of the Hong Kong Special Administrative Region shall continue to recognize the professions and the professional organizations recognized prior to the establishment of the Region, and these organizations may, on their own, assess and confer professional qualifications.

The Government of the Hong Kong Special Administrative Region may, as required by developments in society and in consultation with the parties concerned, recognize new professions and professional organizations.

Article 149

Non-governmental organizations in fields such as education, science, technology, culture, art, sports, the professions, medicine and health, labour, social welfare and social work as well as religious organizations in the Hong Kong Special Administrative

> *Region may maintain and develop relations with their counterparts in foreign countries and regions and with relevant international organizations. They may, as required, use the name "Hong Kong, China" in the relevant activities.*

The principle of professional autonomy was cast in concrete.

To the medical profession, there was one more hurdle to overcome – the training and recognition of specialists. Hong Kong had been a British colony since 1841, and our medical training and health care services were all modelled on the British system. Doctors were trained to be an all-rounder. When they graduated, they were equipped with the knowledge to treat all conditions: from surgical procedures like appendectomy to delivering babies. That means they were supposed to treat patients, not just their diseases.

Yet medicine is getting more and more complex, and the science of medicine is making progress by leaps and bounds. No one can be a master of all. Specialization in one area or another is not only necessary but inevitable.

But how does a medical graduate become a specialist? With structured training and more stringent accreditation! After all, the public wants the best possible service, and the profession is determined to provide the best.

Ironically, Hong Kong did not have a structured medical training programme for any medical specialty before the

1990's. After graduation, a doctor would take up, usually in the department of a public hospital, a job of his choice of specialty for 2-3 years. When he gathered enough knowledge and experience, he would leave for the United Kingdom or any other Commonwealth country to sit for one of those Royal Colleges specialty examinations. If he passed, he became Fellow/Member of that Royal College specialty and returned to Hong Kong as a "consultant" surgeon, physician, obstetrician, pediatrician, radiologist, etc. Ironically, in those days, even the various Royal Colleges did not require applicants for fellowship or membership to receive any structured training.

It was under such circumstances that I became a qualified "surgeon" in early 1965, barely three years after graduation from the medical school of the Hong Kong University in July 1962. By then I was a Fellow of the Royal Colleges of Surgeons of Edinburgh and England, having succeeded in the fellowship examinations of the two colleges.

The imminent change of sovereignty woke us up. We were most concerned with three areas:

- After its return to Chinese sovereignty, Hong Kong will cease to be a British colony and therefore can no longer rely on the United Kingdom as a role model and accreditation authority for its professional bodies;
- We need to develop our own standard training programmes and accreditation procedures to ensure that our "specialists" are in par with international standards;

- The principle of profession autonomy must be upheld.

This meant we needed to design not only a standard post-graduate training programme for doctors, but also an institution that implemented such programmes leading to the registration of specialists.

The formation of the Academy of Medicine

The initial thought was to set up colleges for various specialties, each responsible for determining its training criteria and accrediting its programmes.

Under such a system, each college would determine its own programme without the interference of another college. The problem with such a system would necessitate the setting up of at least 30 different colleges, which would make the formalization of standards extremely difficult if not impossible. This would also be an administrative nightmare and a hard sell to the public.

Rationality won the day in the end. It was agreed that while there should be colleges, there should also be an umbrella body to oversee these colleges and determine the general guidelines on standards. The idea of the Academy was thus conceived.

It was also agreed that:

- it should be a statutory body;
- since it will be a body to set and determine the standards for the medical profession, it must be under the control of the medical and dental professions, upholding the principle of professional autonomy.

The Hong Kong Academy of Medicine Ordinance (Cap. 419) was passed into Law on June 25, 1992 and came into effect less than two months later on August 1. During the second reading of the Bill, I, as an elected member of the Legislative Council representing the Medical and Dental Professions, made the following statement:

> *The formation of the Academy of Medicine, Mr. Deputy President, thus heralds the establishment of our own training programmes through the different academy colleges as stipulated in the Bill. This statutory body, the Academy of Medicine, will also be responsible for the vetting and subsequent accreditation of our own post-graduate status. Mr. Deputy President, this is a move the medical and dental professions yearn for and very much welcome.*

According to the Bill, the Academy will be established by statute and more importantly its five principle officers (President, Vice President (General Affairs), Vice President (Education), Hon. Secretary and Hon. Editor) will be elected within the medical profession and the rest of the Academy's Council members are all *ex officio* members from the various colleges (all members of the medical profession). Professional autonomy was therefore guaranteed.

A *bona fide* medical professional who has the right to practise in Hong Kong and who had received the required training determined by a College will have to go through an assessment process, and if he passes, he will be conferred a Fellowship of that College. Then he will be recommended to the Academy Council for consideration to be made a Fellow of the Academy.

Consequent to the setting up of the Academy, important amendments were made to the Medical Registration Ordinance to add a new category of registration – Specialist Registration. To be registered as a specialist, a medical practitioner must be a Fellow of the Academy or receive the recommendation of the Academy based on the completion of recognized or equivalent training standards.

The setting up of the Academy of Medicine in essence guarantees that every medical graduate has achieved a certain standard of training and passed stringently approved accreditation. To the public, the specialist registration is a mark of confidence. Today, the Academy has 15 colleges under its wings all housed in the Hong Kong Academy of Medicine Jockey Club Building built on a piece of land acquired through the Government and financed by donations by the Royal Hong Kong Jockey Club and philanthropists such as Sir Ellis Kadoorie, the honourable Li Ka-shing (李嘉誠), Sir Run Run Shaw (邵逸夫), Sir Lee Quo-wei (利國偉) among others. The Fellows also played their part in raising money through a highly publicized classic Chinese drama performed in the English language. The first Executive Council consisted of Professor Sir David Todd (達安輝) as President and myself as Vice President (General Affairs). For the first time in Hong Kong, the medical and dental professions were given the

The Hong Kong Academy of Medicine Building (Courtesy of the Hong Kong Academy of Medicine)

right to determine specialist training standards and accreditation to upgrade their service to the public.

Since 1985, the members of the Legislative Council have been elected either via direct election on a geographical area or from members of a profession or body (Functional Constituency). The latter is often snubbed as "small circle election" and the elected members are always sneered at as working for the good and benefit of their professions or bodies only. Small circle election may well be, yet like all elected legislators, they perform their roles in monitoring Government's policies and performance, and push for ways and means to improve the public good. My work as an elected member of functional constituency in the Legislative Council is a case in point. Yes, I answer to the call of my constituent (doctors and dentists). Yet the drive is for public good. The amendment of the Medical Registration Ordinance and the establishment of the Academy of Medicine result not only in guaranteeing the standards of doctors in their practice but also empowering the public to monitor their performance. The truth is whether the Functional Constituency system works depends on the vision and action of the elected representatives, not the Functional Constituency. In other words, it is the "singer" not the "song" that matters. If they do their job properly or "play their cards" right, they will help the Government develops the professions which they represent. If it is the medical profession they represent, they should assist the Secretary of Health to improve the health care services of society. As their representatives, they of course are obliged to listen to the views of their functional constituencies and relate them to the Legislative Council and the public. Yet they are not their voters' mouthpieces. Instead, they owe them their good sense and independent judgment. And as a legislator, they should also work for the collective interests of society.

Chapter 4

Transforming the health care services

Hong Kong runs a dual track health care system – the private service and the public service. The private service, as expected, charges a fee for service. Doctors in the private sector charge a fee for their usually competent services. For the patients, the attraction lies in the range of choices they are given, and a level of comfort and services commensurate with their affordability. They always have a choice, in everything from who are going to treat them to how and when they will be treated.

In sharp contrast, the public health care system is totally subsidized by money from the public coffers. What citizens pay is actually no more than a token, 120 dollars per day since 2018 for basically all services, including all necessary investigations and treatment. Such affordable and comprehensive services are made available to all. The Government's basic health care policy is that "nobody should be devoid of proper care because of lack of means". Priority is determined by the urgency and severity of the patient's conditions. The public hospital opens its door to and it serves as a final resort and safety net for those who either fail to get the treatment they want in the private sector, or simply can't afford it.

It is therefore only natural that the demand for public service has always been enormous. Yes, the Government does have a sizable budget for health care, and this, in money terms, increases every year. Still, this is using a finite budget to satisfy an insatiable demand. While patients cannot and will not be turned away, they have to be put on a queue. And their choices are limited. They cannot choose their service providers because they are cared for by a team. Nor can they choose the time to receive treatment.

Anyway, the services provided by our public hospitals should have won the gratitude of the citizens. But this was not to be, patients on the contrary were often unhappy and dissatisfied – the wait time to see the doctor was long and to be admitted for treatment longer. Hospital environment was far from satisfactory; the wards were so overcrowded that many had to sleep in camp beds. Staff attitude left much to be desired. The staff was unhappy too. They had to work under tremendous pressure and, given the small number of senior positions, their promotion prospects were grim. Training was abysmal as everyone was simply a working horse. All these, coupled with the attraction of the private market where much more money can be made, resulted in an alarming staff turnover rate. In 1988, the average turnover rate for public hospitals was some 18-23%. The Government was frustrated – for all the money it spent; it got only continuous criticism.

The grievances and dissatisfaction of doctors working in public hospitals reached a tipping point in 1988 just prior to the second Legislative Council election. Led by E.K. Yeoh (楊永強), Ko Wing-man (高永文), Ho Shiu-wei (何兆煒) and Fung Hong (馮康), a mass rally was organized and participants threatened to go on strike or to work to rule. Ironically, these activists and "rebels" subsequently joined the establishment and became "preachers".

E.K. Yeoh became the first Chief Executive of the Hospital Authority and later the Secretary for Health and Welfare; Ko Wing-man joined the senior rank of the Hospital Authority and subsequently also the Secretary for Food and Health. Ho Shiu-wai became the second Chief Executive of the Hospital Authority; Fung Hong joined the senior ranks of the Hospital Authority. Under such circumstances, I campaigned as a Legislative Council candidate for health care reform. I was also against any "industrial action" by medical practitioners. It is bad for the reputation of our profession, and it will hurt our patients. For what I believed in and stood for, I won a seat in the Legislative Council.

The fact is obvious. A health care service that is run in a civil service style lacks flexibility. It is bureaucratic, and has little concern for cost efficiency.

The Hospital Authority

The Government by that time was all ears. At a breakfast meeting on one Saturday morning at the coffee shop of the now non-existent Furama Hotel in Central, the Colonial Secretary David Ford met the Secretary for Health and Welfare Mr. Brian Chau (周德熙) and myself. The Government agreed on the following:

With Sir S. Y. Chung (centre) and members of the first Hospital Authority Board (Courtesy of the Hospital Authority)

- Overnight there will be an increase in the number of senior posts in public hospitals (consultants and Senior Medical Officer) ;
- Adopt the principle of reform as proposed by the Australian Consultants (Coopers and Lybrand) to take the public hospital services out of the civil service and to set up a body to oversee the implementation – the Provisional Hospital Authority.

Sir S.Y. Chung (鍾士元) was appointed to chair the Provision Hospital Authority and, as an elected representative of the Medical Functional Constituency, I became a *de facto* member. Our job was to set up a structure for the new body, determine how the different hospitals should function, engage the staff properly, talk with the two medical schools and arrange a system for the allocation of funds from the Government, and ensure that the reforms would eventually serve the best interests of both the medical staff and the Hong Kong people.

Sir S.Y. Chung proved to be a legend. A stalwart politician who is just as well-versed in the art of compromise as in the knack for getting things done. He is by profession an engineer and an extremely seasoned manager. As both an appointed member of the Legislative Council for many years and a senior member of the Executive Council, he had served many Governors. By the time the idea of a Hospital Authority was conceived, he was in his early 70s. When he was initially approached by the Government to head the Provisional Hospital Authority, he said no, citing his old age. David Wilson, the then Governor, refused to give up. After repeated attempts, he persuaded Sir S.Y. Chung to accept the appointment. Sir S.Y. Chung pledged to do his best as it would be his "last job".

He did exactly that. Sir S.Y. Chung was still as sharp as a hawk. He was determined to turn the hospital service upside down from the bureaucratic Medical and Health Department. As the Chairman, he was extremely hands-on and often played the role of the Chief Executive. He threw out ideas and pushed them through with his clarity of vision and intensity of commitment. He scrutinized every policy paper submitted for discussion and read the minutes of every meeting. He solicited members' views and argue with them. Most of the time, he turned out to be right. He was also dedicated, attending every meeting and never failed to be on time. It was because of his support that we won the Government's approval for the funding policy of the Hospital Authority. It was also because of his perseverance that the Government finally consented to encash the Flexible Spending Account for the staff. The Hospital Authority owed much to his visionary leadership and his rapport with the Government.

Many have thought, wrongly, that the formation of the Hospital Authority was an attempt of the Government to overhaul the health care system. It was not. Setting up the Hospital Authority was in essence a public hospital management reform. The concept is to remove all public hospital services (both the Government Hospitals and subvented hospitals run by charitable organizations) from the civil service, and place them in the hands of a statutory body on a fixed annual budget funded wholly by the Government. It was only much later that the outpatient service was also entrusted to the Hospital Authority.

It would be foolhardy to believe that transformation of such a magnitude would be easy. While the concept seemed agreeable and its implementation may provide better public services, the "devil" as always, was in the details.

Yet we have to soldier on and climb every mountain, so to speak. A quick look at some of the vexing issues we had to wrestle with will give an idea of how challenging our task was:

- **On what principle was the Hospital Authority to be budgeted by the Government?**

It was initially suggested that the budget allocated to the Hospital Authority will *not* be more than if the Hospital Authority did not exist. This was obvious unacceptable, for "*not* more than" implied that it could be less. My counter offer was the budget "will not be less than if the Hospital Authority did not exist." Ultimately, prudence won the day and the final principle was "that the budget for the Hospital Authority will be *comparable* to the one if the Hospital Authority did not exist".

- **What new package will be offered to staff?**

One of the reasons for the formation of the Hospital Authority was that there was an employee exodus, in particular the medical staff. The Hospital Authority, therefore, is set up partly to "attract, retain and motivate" staff. Yet operating on a shoestring budget with no extra funding, how could we pull this off? An analysis of staff emoluments revealed that their remuneration packages consisted of not only a salary component, but a welfare component. In the case of a consultant medical officer, the welfare portion actually amounted to some 110% of his salary. If we could encash the non-essential fraction of his welfare, like overseas education allowance for his children, or non-departmental quarters, his actual take-home pay could be raised dramatically. For the consultant doctors, this amount to 60% of their basic salary (Flexible Spending Account). That means a consultant's take-home pay was increased 160% overnight. This would effectively curtail the exodus.

- **How do we deal with staff movements?**

There were two categories of staff in the public hospitals. Those who worked in the Government hospitals such as Queen Elizabeth Hospital or Queen Mary Hospital were civil servants entitled to the benefits of civil servants including pension. Those employed in subvented hospitals – run by charitable organizations with financial grants by the Government such as Tung Wah, Caritas, Yan Chai, Buddhist, Lady of Maryknoll, etc. – were employees of the charitable organizations.

Given these differences, would they accept the Hospital Authority terms and condition of services? Naturally those from subvented hospitals were more willing. So were the more junior staff in the Government hospitals who found the higher take-home pay attractive. Yet there were staff members who would not be willing to give up their pension, especially senior staff on the verge of retirement. Finally, a compromise was reached – all new recruits would be employed on the Hospital Authority terms. Existing staff were given a choice. Those who chose to remain as civil servants were allowed to do so until retirement. A Government Hospital Service Department was created to take care of them. They would continue to enjoy all the benefits while under secondment to the Hospital Authority.

- **How to deal with the "plight" of subvented hospitals**

What about the subvented hospitals? Hitherto they were under the management of the Board of Directors of the charitable organizations. These Boards employed the entire staff, and the hospital management reported directly to the Boards.

Financially, the hospitals received grants from the Government via the Director of Medical and Health Service. Now with the establishment of the Hospital Authority, the Government would fund these hospitals in the same way as government hospitals and the hospitals' superintendents (Hospital Chief Executives) would report to the Chief Executive of the Hospital Authority. It was therefore an arrangement of give and take. Again, rationality prevailed at the end of the day. The charitable organizations' Boards of Directors would still determine policy directions of their hospitals as members of the Hospital Governing Committees. That means the mission – customs and cultural practices of these charitable organizations would be preserved at their hospitals. The management of these hospitals fell within the ambit of the Hospital Authority.

Should all public hospitals be transformed at the same time? Singapore, which was reforming the management of its public hospitals at around the same time as Hong Kong, took a more conservative approach by transforming their hospitals one by one. Under the leadership of Sir S.Y. Chung, we, however, decided to act boldly – by introducing the new management scheme into public and subvented hospitals all at once! The transformation was a mammoth task for it involved over 44 hospitals and 53,000 staff members.

Having addressed the major concerns and worries of the various stakeholders, the Government moved to introduce the Hospital Authority Ordinance which was passed in 1990. The Hospital Authority became a statutory body and members of the first Authority Board were appointed, again with Sir S.Y. Chung as chairman and myself as member. We were now all set to put our ideals into practice.

In short, the Hospital Authority introduced four initiatives:

- Having come out of the civil service, it will discard the civil service bureaucracy;
- The service of public hospitals will be provided based on a cost-effective principle through proper management;
- There will be public participation in the policy formulation and management of the public hospitals;
- Consultant doctors and senior nurses are now entrusted with management responsibilities. They will be budget holders answerable to performance pledges. Senior staff is now given a dual career ladder – as healthcare professional or manager in hospitals. This would ultimately lead to a surge in the need of health care managers.

They proved to be game-changers. The public hospital system came out of the doldrums. Staff morale was revitalized and public complaints decreased. We introduced public participation in formulating the management policy of the Hospital Authority through appointed members, regional leaders and members of the public with specific expertise on the Board of the Hospital Authority, the Regional Advisory Committee and the Boards of the different hospitals. An increasing number of patients were lured back from private to public hospitals for their money-for-value services. To such an extent that public hospitals became a "victim of its own success", they are having more patients than they could serve.

The Hospital Authority development had also brought another windfall for health care services. Hitherto, Hong Kong private hospitals were "in hibernation", resting comfortably in their comfort zone. Now facing increased competition from the public sector, they shook off their complacency and started to improve their services and facilities.

Hong Kong has one of the best public health care systems in the world. It is affordable and accessible to all and everyone is treated equally. Its underlying principle for public health care is: "Nobody should be deprived of care (health care) because of his lack of means". Currently, inpatients pay a token of 120 dollars per day irrespective of the investigations and treatment they receive. The elderly and those on welfare are partially or totally exempted.

Unfortunately, a seemingly philanthropic system like this is also subject to abuse, as everyone, irrespective of means, is given access to a complete range of heavily subsidized and high-quality health care services. It is also a black hole for the Government using a limited budget to supply an ever increasing demand.

Going down this philanthropic road, the standard of Hong Kong's health care system, the pride of its people, are destined for a slow but inevitable decline. A thorough and critical examination of the principle of public hospital and health care financing is in order. The failures were obvious. The Hong Kong people are so used to a medical service which is efficient, accessible to all, in par with world standard and next-to-totally-free. Any revamp of health care finance for public medical service would entail money taken out of one's own pocket.

For some 20 years, there have been proposals after proposals to remedy the situation, but so far none has gone off the drawing board. More recently, for example, it is suggested that those who can afford it should be encouraged to take out medical insurance. With insurance coverage, these people can then seek the service of the private sector, thus redressing the imbalance between the public and private health care sectors.

Then again, the devil is in the details. For this to work, many problems need to be solved. In particular a compromise needs to be reached between the insurance industry and the medical sector. The insurance industry may be providing a service to the public, but it is primarily concerned with profits. The medical profession, however, wants only the best for the patients. In short, the two have vastly different priorities.

If rationality is to triumph, a few basic principles have to be agreed upon:

- Health care is an essential service whose provision no responsible Government should run away from;
- Money from the public coffers should be used for those who need it most.

In accordance with these two principles, should our health care system be defined as "a safety net" and "those who can pay should pay or at least co-pay"? Is there a place in the system for "target subsidy"?

Let me elaborate. I believe very few people would object to providing maximum subsidies for the old and the destitute even for the treatment of minor ailments like a common cold, to put into practice the motto that "nobody should be devoid of care because of lack of means".

But it is an entirely different matter to subsidize everyone for minor ailment, including those who can afford it. It is only when they have a major illness that requires long-term treatment or state-of-the-art technology that may dry up their savings should they be subsidized. In other words, the degree of subsidy should reflect the financial status and the financial burden of the illness that the patient has to face. A system of co-payment needs to be put in place.

I had been proposing this idea for many years, yet it had never been taken seriously for the simple reason that it was considered difficult to implement and administer. To me, nothing is difficult or impossible if it will help the situation. It is a matter of political will and determination.

In and around 2000, when, as a member of a delegation from Hong Kong to visit Singapore to take stock of its health care system under the leadership of the then Secretary for Health and Welfare, Mr. Brian Chau (周德熙), I had a private audience with the country's Minister of Health. I came away from the meeting with this understanding: "In Singapore, nothing is free, everything is charged according to means. In Hong Kong everything is free or presumed to be free." Here lies the difference. Singapore is using the basic concept of "target subsidy".

Reform in the A&E departments

At the public hospitals of Hong Kong, the Accident & Emergency (A&E) Departments are always overcrowded, especially on weekends and public holidays. The situation gets worse by the year. Surveys showed that as many as 60% of the "patients" attending the departments are in no need of emergency care. Many frequent the A&E for treatment of common cold, stomachaches, headaches, insomnia, etc. A&E Departments are therefore seriously abused. The reasons are simple:

- A&E Departments are opened 24 hours a day (7 days a week) and on public holidays;
- It is totally free;
- It provides a one stop service – consultations, simple investigations (blood and urine tests and simple X-rays) and possible treatment;
- Few private general practitioners provide service in the evenings, on weekends and the public holidays.

A & E Departments, therefore, become a safe haven for people even with simple conditions seeking consultation at their leisure.

I remember leading a group of legislators on a visit to the A&E Department of Pamela Youde Nethersole Eastern Hospital around 8pm after dinner. The Department was bursting to its seams. A new patient came in to register. He was given a registration number. "You probably have to wait for three hours," a kind nurse told him. "Doesn't matter, actually I have a mild

An overcrowded A&E department (Courtesy of the Hospital Authority)

headache and I couldn't sleep well so I thought I would walk in and see a doctor. Oh, incidentally I have to visit a friend in your surgical ward. This is my mobile phone number. Call me when it's my turn."

It was a blatant abuse of the A&E services, it also put unnecessary extra burden on to the staff of an already extremely overloaded department. Yet who could blame him? He had an ailment that he needed advice and there was no place that he could turn to!

I made a bold decision. In 2002, when I took the helm of the Hospital Authority, I introduced a charging policy to the A&E Department for all attendances who were deemed non-emergencies. It was a daring move. Overnight, a time-honoured non-charging policy that had been in force since a very long time

Announcing new charging policy with Ho Siu-wai (Courtesy of the Hospital Authority)

ago was scrapped. Needless to say, the elderly and those receiving security allowance would be exempted. My "opening bid" to the Government was 200 dollars per visit, the Government got cold feet. Finally we agreed on 100 dollars. I was confident I would prevail. The charges were not meant to cover cost, but to weed out abuses. On the first day when the charge of 100 dollars was introduced, there were no objections, no demonstrations, not even a placard to produce a fanfare. It was all quiet on the Western Front!

Regrettably, the anti-abuse effect was shortlived. With no other facilities to go to, most citizens were quite willing to fork out an extra 100 dollars.

Chapter 5

Are leaders born?

You put your friend on a pedestal by complimenting him that "he is a born leader". But leaders are *not* born. They are people who rise to the occasion, impress people with their ability to make rational decisions and ultimately "bite the bullet" when the going gets tough.

When I was appointed to manage a health care system of some 50,000+ staff and 40+ hospitals, I was running a medical practice with two nurses and a secretary on my payroll. It was a "big leap forward" for me. The job was relatively easy during "peace time" – you carry on the routines, make public appearances, officiate at hospitals' public functions, say things that the Government like to hear and you will be praised and saluted. But it becomes an entirely different ball game when you have to deal with a full-blown crisis during "war time". You have to make bold, difficult decisions under tremendous pressure. As a leader, you have the obligation to protect the good name of your organization and the best interests of your people, and at the same time you are responsible for the health of the citizens. You want them to emerge from the crisis even stronger than before.

The year 2003 began unremarkably. Implementation of the

new charging policy in the A&E Departments went without a hitch. Like the rest of the developed world, Hong Kong was oblivious to the sudden outbreak of epidemics and infectious diseases. Yes, we still have endemic pulmonary and extrapulmonary tuberculosis; cholera and typhoid, etc. But Hepatitis B was under control via inoculation to every new born child, HIV/AIDS infection had never been alarming due to effective education campaigns and cross-border control. Again, like the rest of the developed world, Hong Kong was focused on lifestyle diseases, non-communicable diseases like malignancies, and a range of health problems associated with an ageing population. There had been an attack of avian flu, but it was controlled by culling all life poultries in infected farms, and forbidding the sale of live poultries.

We were so complacent that there were only a handful of isolation beds for infection diseases at the Princess Margaret Hospital and nowhere else. Perhaps not surprisingly, there were only a few doctors who took up the specialty of infectious diseases in 2003.

Yet unbeknownst to all, infectious diseases were always doing their best to make a comeback. I spoke at length on this topic in the McFadzean Oration delivered on October 23, 2004 as the President of Hong Kong Academy of Medicine (See Appendix 1 for the full text of the speech) .

In the early spring of the same year, there were reports (subsequently proved to be true) that people in South China were rushing to buy "white vinegar" to "sterilize" homes to prevent an unknown respiratory tract infection. While the local health authorities in Hong Kong took note of this, no precautionary measures were taken and there was still no inspection at the border. In April of the same year, there was a meeting of the world's leading health organizations in Hong Kong attended by such luminaries as the Head of the World Health Organization (WHO) Western Pacific, the Minister of Health from the Central Chinese Government (CCG) , the Director of Health of the HKSAR, and the Secretary of Health and Welfare of the HKSAR. It was held at the Hong Kong Academy of Medicine Building. The Chief Executive of the Hospital Authority and myself as chairman of the Hospital Authority were also in attendance. As the meeting closed, the Minister of Health of CCG briefed the local media, saying that despite isolated cases of unknown respiratory infections in the Mainland, there was no cause for alarm.

Around the same time, a medical professor arrived secretly in Hong Kong from South China and checked into the ill-fated Metropole Hotel apparently to seek treatment for the unknown respiratory syndrome. That started the whole saga that was to last 100 days in Hong Kong. We went into a total economic collapse; a near paralysis of our tourism industry, both inbound and outbound – Hong Kong was put on the traveller's alert list, our students returning to US and UK to continue their studies need to be quarantined; an international annual watch and clock trade show in Geneva barred Hong Kong's participation. Before the outbreak was finally over, it afflicted 1,755 people, and took away 299 lives.

Visiting hospital staff during SARS as a routine (Courtesy of the Hospital Authority)

What made this chapter of Hong Kong history even more tragic was that relatives of the victims were not allowed to have contact with their love ones who were dying of Severe Acute Respiratory Syndrome (SARS) for fear of contact of the disease. For the same reason, most funeral parlours refused to provide services. The city had never looked more desolate. Everyone was faceless, covered by masks of varying color, shapes and sizes. Everywhere I went, I saw people keeping their distance and avoiding one another. I personally felt like a leper in the "dark ages". The public sprang away from me to keep a distance. It was melodramatic that late at night one day, I went with my wife to patronize the coffee shop of the Grand Hyatt Hotel, having had no food for the whole day. The coffee shop was usually well patronized on good days. That night, however, there was only a handful occupying two tables. As we were being led to our table, the customers at the other two tables got on their feet and moved to the tables further away from us. I instantly understood – I, as the chairman of the Hospital Authority, was possibly unclean and they were not taking any chances.

This nightmare, however, brought out the best in Hong Kong people. Society became once again cohesive – protective garments poured in, along with well wishes and gifts of fruits and herbal medicine. An endless stream of all sorts of Chinese desserts were delivered to the public hospitals to support our dedicated staff from all walks of life and foreign consulates, non-Chinese communities, business sectors and labour unions. In return, health care professionals gave their very best. None walked away from their responsibilities – there were *no* deserters, only a blatant display of comradeship, courage and self-sacrifice!

SARS appeared to have a "liking" for health care workers, especially the younger ones. Every morning when we met in the "war cabinet" at the Hospital Authority headquarters there were reported cases of staff succumbing to SARS. Taking up the responsibilities of looking after SARS patients in the designate SARS wards was therefore like being sent off to purgatory, you could be signing your own death warrants. During those days, you could often hear this from one health care worker to another at public hospitals: "I will take over from you. You have a young child/elderly parent to take care of, I have no major family ties, and my wife has her own job." Many volunteered to work in the SARS ward, knowing all too well that chances were they could be the next victims. Many were reluctant to go home for fear of infecting their family members. As a result, the Hospital

Authority had to provide temporary dormitories for them and many didn't get to see their love ones for days or even weeks.

So the entire Hospital Authority staff was groping in the dark and fighting a totally unknown disease. Every day, they saw their brothers and sisters in arms fall prey to the deadly disease. At the same time, "phone in" radio programmes every morning were pointing accusing fingers at us. This was unfair and frustrating but we stood our grounds and soldiered on. While the frontline staff was immersed in a "war without smoke" with SARS, the management had to fight a total different battle.

At that time, my most urgent tasks at hand were:

- Estimate the possible caseload, the availability of hardware (ventilators, protective garments, etc.) and software (staff) and where and how to accommodate the patients;
- How to maintain the routine hospital services. After all, the Hospital Authority, being the ultimate "safety net" in Hong Kong's health care system, had to continue to provide the routine services, emergency services and handle referrals from private hospitals of suspected SARS cases. SARS was an infectious disease that must be treated with extra care. Meanwhile, I was highly vigilant about:
  - public expectations;
  - Government expectations;
  - staff morale.

It would not be possible to estimate the number of victims. What we could tell with confidence, based on the amount of hardware and software required for treatment, was the maximum number of patients we could deal with or accommodate. Our estimate was around 3,000, which was the number of patients for whom we have facilities (ventilators, oxygen tent, adequately trained frontline staff). I believe sharing this knowledge with Hong Kong people would give them confidence and help them sleep better at night.

Regrettably, this did not go down well with the top level of the Government. When I expressed the view on air at the Radio Television Hong Kong (RTHK) programme, *Newsline*, I was given the "marching order". To them, this was tantamount to saying that Hong Kong would have some 3,000 victims and this message would damage Hong Kong's economy.

As more cases were admitted to public hospitals, a decision had to be made on whether we should designate a few hospitals as SARS hospitals, or let patients scatter throughout the current 13 hospitals with acute A&E admissions.

After discussing with my senior management team, I decided to put most SARS victims in Prince of Wales Hospital (PWH) and Princess Margaret Hospital (PMH). PWH had already housed most of the initial SARS victims and PMH was Hong Kong's only designated infection disease hospital. It was also close to Amoy Gardens – where the first outbreak of SARS happened.

Then we staged a massive decanting operation overnight – PMH's surgical and non-emergency patients were transferred to the Caritas Hospital. My gratitude to the staff concerned.

Was that a wise decision? I believe that it was, at least it was a rational one. After all, in the demand for routine and

Briefing Chief Executive Tung Chee-hwa (centre) on SARS

emergency services provided by public hospitals didn't abate with the outbreak of SARS. That decision, however, was thoroughly and brutality chastised by the Health Panel of the Legislative Council. This was most unfair and uncalled for. Again, while we were emptying our tanks, our honourable legislators were finding fault with us in their air-conditioned offices. None of them had shown any genuine concern for our "fighters", none had visited the hospitals to boast staff morale. The Chief Executive of a public hospital was entirely justified when she asked them, "Where were you, when we needed you most?"

One senior Government official challenged me on why there had been so many on our staff falling prey to the disease while "there wasn't any victim dying of the disease" in Guangzhou, implying that we had not done enough and had not been vigilant enough. I didn't bother to respond, except to say that SARS was brought to Hong Kong by a medical professor who "slipped" into the city from Guangzhou to seek treatment.

In a battle such as this, staff morale was the most important. It could rise and fall like a yo-yo. The morale of the Hospital Authority's staff had dealt a heavy blow when its Chief Executive Ho Shiu-wei was down with the disease. Ko Wing-man, acting as Chief Executive and myself were unperturbed. Though I was quite sure that both of us did have a small dose of the infection, we stayed at the forefront and continued the routine of visiting and comforting staff of various hospitals, and making public announcements on the status of the epidemic, and taking measures to ensure that adequate protective gears were distributed.

For 100 days, I met with my senior management team every morning at the headquarters of the Hospital Authority to:

- receive the report and analysis of the SARS statistics of the previous day;
- discuss and make decisions concerning the implementation of policy and the need to make policy changes according to circumstances;
- discuss and plan the routine work of the hospitals.

As the acting Chief Executive, Ko Wing-man had to act as a bridge between the Hospital Authority and the Government, and had to attend Government meetings every day. The public was concerned and would like to know the actual situation. A press

Delivering eulogies to staff lost to SARS

briefing was arranged every afternoon. In spite of this, both Ko Wing-man and myself were hounded by the morning "phone-in" programmes. As the Chief Executive, Ko Wing-man took all the unfair and abusive assault with dignity and protected the integrity of the Hospital Authority. To Ko Wing-man, I am forever grateful.

How could I be so calm at the eye of the storm? We, the senior management and the frontline staff, walked shoulder to shoulder for the same goal and purpose. The decisions we made might be right or wrong, but they were all made in good faith after detailed discussion and therefore with no regrets. The commitment of the whole team, not shaken by adversity, gave me a better understanding of what dedication is all about. It also demonstrated the beauty of human nature.

I fell into profound grief every time I delivered eulogies for our staff lost to SARS, not just one but six. Yet I was also inspired by their noble sacrifice and I was proud not only of the deceased but also the entire staff.

In 2003, when the outbreak finally came to an end and Hong Kong was moving back to normalcy, I accompanied the then Premier of China – Wen Jiabao (溫家寶) on a visit to Amoy Gardens and the Prince of Wales Hospital, the Premier mingled with the staff of the Hospital Authority, rolled up his sleeves and shook their hands. It was an enormous shot in the arm for the Hospital Authority. I was moved. This was what leadership is all about!

Apologies had to be offered to our patients and their relatives. Once they were admitted and confirmed to have SARS, they were banned from seeing their love ones. Sadly, some passed away without a chance to express their last wishes. This was a difficult decision which had been made for the protection of public health and the prevention of the spread of the disease.

The end of SARS was not so much a medical triumph as a vindication of well tried-out public health principles – detection of the infected persons, isolation, and quarantine of their contacts. These procedures were carried out by the Home Affairs Bureau. And the tracking of the patients' contacts was a joint effort of the Hospital Authority and the Hong Kong Police Force. To them I salute!

Towards the end of May, as we could see that the number of SARS cases was falling, we, at the Hospital Authority, began

Premier Wen Jiabao celebrating SARS heroes

His autograph paying tribute to Hong Kong medical professionals

to take a critical look at what we had done to combat the crisis, deliberately looked for shortcomings and make recommendations for the future should another "epidemic" befall. It was not that we thought we had done anything wrong. We believed it was the responsible thing to do.

The Government, on her part, lost no time in starting her own "independent" investigation. But since the investigation was headed by the Secretary of Health and Welfare, it was widely criticized by the public as "self-investigation".

The legislature seized the opportunity to set up a "Select Committee". That was a "witch hunt" designed to "draw blood", and it would stop at nothing until "heads began to roll". Still, the Hospital Authority did its best to cooperate with precious staff time and financial commitment. I decided then that should the report find any fault with the staff or the Hospital Authority's management, I would take the political decision to resign, which I did, on the eve of the publication of their report.

I resigned not to abdicate my responsibility, but to take the political responsibility as the head of a statutory organization, which was a political appointment. I also believe that:

- the Hospital Authority had done nothing wrong. Yes, we had bitter experience and we did make some tough and heart-wrenching decisions. There was nothing for me to apologize;
- it is time to stop the reopening of "old wounds" of those who suffered the disease, and the relatives of those who passed away;

Announcing my resignation (Courtesy of *Sing Tao Daily*)

- I had to put an end to the witch hunt on the medical staff, so that the Hospital Authority would not be demoralized and distracted from its daily routine and preparing itself for future "epidemics";
- the working routine of the Hospital Authority to serve the public and its development could be dragged by unnecessary investigations.

The final chapter of SARS was thus a melodrama and no person won. Yet it did have a silver lining: it showed that when Hong Kong pulls herself together, she can not only survive a serious crisis, but comes out of it a winner. It showed that Hong Kong people can work closely together in face of a major challenge. We made new commitments to further develop infectious disease protocols and to build a new infectious

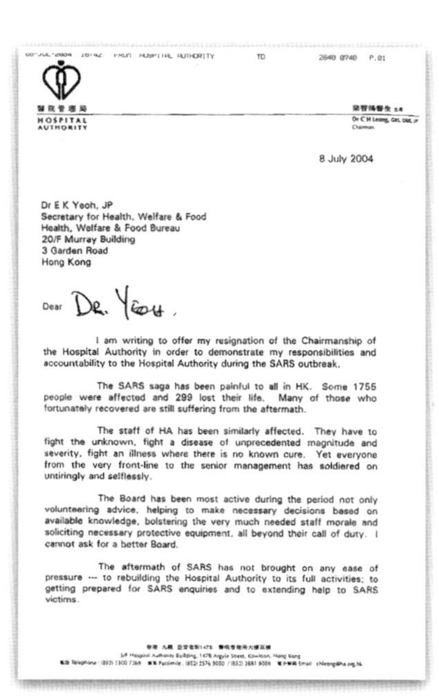

HOSPITAL AUTHORITY

Dr C H Leong
Chairman

8 July 2004

Dr E K Yeoh, JP
Secretary for Health, Welfare & Food
Health, Welfare & Food Bureau
20/F Murray Building
3 Garden Road
Hong Kong

Dear Dr. Yeoh,

I am writing to offer my resignation of the Chairmanship of the Hospital Authority in order to demonstrate my responsibilities and accountability to the Hospital Authority during the SARS outbreak.

The SARS saga has been painful to all in HK. Some 1755 people were affected and 299 lost their life. Many of those who fortunately recovered are still suffering from the aftermath.

The staff of HA has been similarly affected. They have to fight the unknown, fight a disease of unprecedented magnitude and severity, fight an illness where there is no known cure. Yet everyone from the very front-line to the senior management has soldiered on untiringly and selflessly.

The Board has been most active during the period not only volunteering advice, helping to make necessary decisions based on available knowledge, bolstering the very much needed staff morale and soliciting necessary protective equipment, all beyond their call of duty. I cannot ask for a better Board.

The aftermath of SARS has not brought on any ease of pressure --- to rebuilding the Hospital Authority to its full activities; to getting prepared for SARS enquiries and to extending help to SARS victims.

Hospital Authority Building, 147B Argyle Street, Kowloon, Hong Kong

-JUL-2004 18:43 FROM HOSPITAL AUTHORITY TO 2840 0748 P.01

It is my hope that with my resignation and assuming all the responsibilities of HA during SARS and the robust actions taken by HA with the Health, Welfare and Food Bureau and Department of Health, the whole painful experience of the saga could be put to rest and that HA can plan ahead for an even better service to the public in the days to come.

It has been an honour working with you at the Bureau, and with the Board and senior executives and the front-line workers in HA. My service in the HA has been the most fruitful years of my career.

Sincerely

(Dr C H Leong)
Chairman
Hospital Authority

TOTAL P.01

My resignation letter

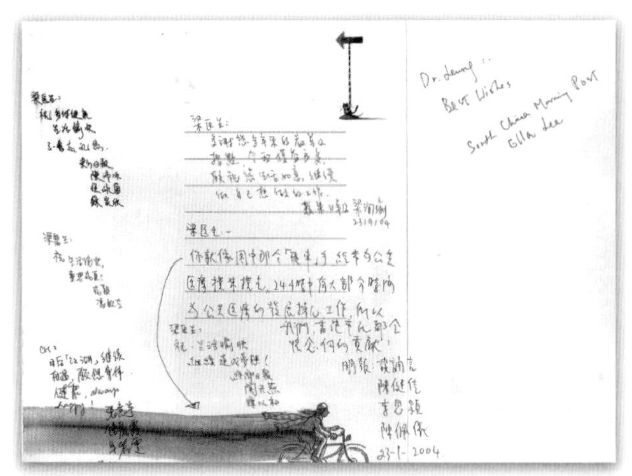

Media friends sent their regards

disease wing next to the old Princess Margaret Hospital. Most importantly, it showed the true colours of the health care professionals in Hong Kong who are willing to risk and give their lives to carry out their duties and obligations to the people they serve.

Chapter 6

Encompassing all – yet giving the best to our citizens

Hong Kong has always been proud that we are an extremely free city. We have no restriction on currency movement. Our people have the freedom to move in and out of Hong Kong. Most other nationals can stay in Hong Kong for a limited period without the need of a visa. In the same way professionals can come to Hong Kong and practise their professions without a working permit as long as they have a license to do so.

The need for a license or to be registered is to ascertain that a professional is up to the standard to ensure the best for our citizens. This is especially important for health care professionals – doctors, dentists and nurses – as human lives are at stake in what they do.

Medical graduate registration examination

Each profession therefore has a "council" which grants registration and issues license based on standards and standards alone. It serves a gate-keeping function that safeguards the interests of citizens.

Under the British colonial rule, the Hong Kong Medical Council mirrors the General Medical Council (GMC) of the

United Kingdom. This Council (GMC) oversees the standards of all the medical schools in the UK and her colonies and the Commonwealth, by assessing their curricula and standards every five years. Any medical school whose standard is approved by the GMC is therefore acceptable to the Hong Kong Medical Council and their medical graduates are granted registration status and given a license to practise medicine in Hong Kong. The GMC also assesses the standard of the medical schools of the University of Hong Kong and the Chinese University of Hong Kong. It is under this arrangement that the medical school graduates of the University of Hong Kong and the Chinese University of Hong Kong are granted their registration status and the license to practise when they graduate.

The GMC's role was a blessing for Hong Kong in the colonial days. In the 1950's and 1960's, Hong Kong was extremely short of medical professionals. Local graduates were less willing to work in subvented hospitals whose facilities were not as advanced as those of the Government hospitals. They also had fewer headcounts for senior positions and therefore fewer opportunities for promotion and career advancement. Local graduates were also reluctant to work in unpopular Government clinics in rural districts, outlying islands, prisons, etc. Graduates from the Commonwealth countries, in particular, Burma (Myanmar) , saved the day. They were graduates from the Commonwealth universities approved by the GMC and therefore entirely qualified to practise in Hong Kong. As a result, subvented hospitals such as

Tung Wah, Yan Chai, Caritas, and unpopular clinics on outlying islands such as Lamma Island, in rural districts, floating clinics (慈航號), even in Stanley prison were all served by Burmese. They had done a fantastic job in filling the void and deserved our gratitude.

But the supply of Burmese doctors was still not enough to make up for the shortfall in general practitioners providing primary health care, especially in the "rural" areas of Hong Kong. In the 1950's and 1960's, newly developed areas such as Kwun Tong and Wong Tai Sin were considered "rural". It has to be remembered that there was then only one medical school (at HKU) with just 30-40 graduates every year. The problem was aggravated in the late 1940's and early 1950's by the regime change and political turmoil in China. During the Cultural Revolution, for example, Hong Kong's population increased many folds by refugees fleeing from the north. Hong Kong needed more frontline medical professionals.

Among the fleeing refugees were medical graduates from China. Unlike medical graduates from the Commonwealth countries, however, they were *not* registrable and therefore would not be granted a license to practise.

In view of the serious shortage of basic medical manpower, the Local Medical Council under the chairmanship of Dr. Gerald Choa (蔡永業), the then Director of Medical and Health Services, arranged a very simple oral examination for these Chinese medical graduates. If they passed, they could take up the post of

Assistant Medical Officers (AMO) in public hospitals under the supervision of a Medical Officer. Otherwise, they could practise, not as a doctor, but as a "clinic licensee" (診療所持牌人) and be employed by a community organization, such as the Chinese Gold and Silver Exchange Society Clinic or Tongxiang Hui (同鄉會) Community Clinic.

This was another forgotten page of the history of medical services in Hong Kong. In trying to make a living by "practising" medicine and providing services to the people of Hong Kong, these medical graduates from China often experienced hardships and frustrations. Community clinics mushroomed during the early 1950's and 1960's. Many were affiliated with political organizations in the Mainland and Taiwan, some were established to provide basic health care services to their members and the working class. They didn't have enough money to put on their payroll graduates from local medical schools who wouldn't be eager to work in these settings anyway.

The status of these medical graduates as clinic licensees was degrading to say the least – they could only see patients in the clinics, and could only prescribe allowed drugs from a list of approved medications. Strictly speaking, they were not "doctors" in their own right. One graduate lamented that, when required to complete a form of personal data, he found it difficult to fill in the column of "occupation".

It was humiliating, but there should be no denying that these graduates provided much-needed basic medical services to millions in areas where Hong Kong graduates "fear to tread".

Some of the more ambitious or resourceful of them would take a further Licentiate Examination. If they passed, they would be registrable and therefore able to practise as a Licentiate of the Medical Council (LMCHK) .

As Hong Kong continued to develop, most of these community clinics had no more roles to play, the medical graduates who worked there became, as I said, a forgotten page of Hong Kong's medical history.

As the year 1997 drew near, it was getting obvious that the Medical Council of Hong Kong could no longer depend on the GMC for the registration of medical graduates who want to practise in Hong Kong. When Hong Kong ceased to be a British colony, it would be politically unacceptable to continue to follow the recommendations of the GMC. Furthermore, there would be graduates from other places who needed to be assessed. Unfortunately, at that time, China, our motherland still did not have a universal or even a uniform national accreditation system.

What "criteria" or "standards" should Hong Kong use to assess non-local medical graduates applying for a licence to practise in the city, so that it would be seen as fair and all-encompassing, yet giving the best protection to our citizens?

The answer, we believe, is an overall licensing examination for non-local graduates that ensures those who pass comply with a basic standard of professionalism.

Under this arrangement, anyone who is a *bona fide* university medical graduate qualified to practise medicine in his/her place of origin, and has taken and passed a simple licensing examination administered by the Medical Council of Hong Kong can register to practise medicine in the city after a supervisional period of assessment in approved hospitals .

Some call this "protectionism" but this is far from the truth. It is a fair system that doesn't take into account the graduates' nationality and political inclinations.

There were others who called for opening the registration system for graduates of certain medical schools. But this could be seen as unfair and discriminatory. And without a proper transparent assessment of each individual, we could be putting our citizens at risk.

Getting the Bill for the Universal Licence Examination for non-local graduates passed was a steep uphill battle. There was strong opposition from our lawmakers and even the Government officials, many of them with offsprings and relatives studying medicine abroad.

Yet the medical profession and the Medical Council stood their ground, steadfast in their belief that it was a fair, inclusive system that serves the best interests of Hong Kong people. Licensing examinations for medical graduates is by no means uncommon in other parts of the world. Hong Kong medical graduates, for example, would have to sit for a licence examination to practise medicine in the Mainland, just as medical graduates from California have to sit for a licence examination to be registered to practise in New York or Arizona.

As usual, rationality won the day. The Bill was passed into law with one proviso: students who had already been admitted to medical schools in universities of Britain and the Commonwealth recognized by the GMC before July 1997 would be given concession to obtain registration when they graduate.

Recognition of traditional Chinese medicine in Hong Kong

Traditional Chinese medicine (TCM) has been practised in China for over 5,000 years to this day. Our forefathers, be they royalties or commoners, had been kept well and properly treated by TCM when they fell ill.

While TCM is regarded by many as alternative medicine, this is not the case in China where TCM has always been the mainstream medical practice. The fact remains that where there are Chinese, TCM will always be very much in demand.

In Hong Kong, while TCM has always been popular, it was never given the recognition it deserves, let alone a push for further development.

As a result of the signing of the "Treaty of Nanking" in 1842 and other unequal treaties, Hong Kong was ceded to the British who exercised control over almost everything in the city, except Chinese Customs and Chinese cultural practice. TCM was considered a Chinese customs, not proper health care. For over 100 years, TCM in Hong Kong was left to its own devices with no official status and regulations. Any Chinese person could start a TCM practice and call himself a traditional Chinese medicine practitioner (TCMP) .

With the imminent change of sovereignty and the promulgation of the Basic Law, a role will therefore be created for TCM as a legal provider of medical and health services.

Article 138 of the Basic Law stated that "the Government of the Hong Kong Special Administrative Region shall, on its own, formulate policies to develop Western and traditional Chinese medicine and to improve medical and health services......"

Yet what makes a "proper" traditional Chinese medicine practitioner? What standards could Hong Kong use to determine his professional status when every Chinese person can practise TCM in Hong Kong?

A general survey revealed then that there were at least 12 registered trade associations whose members were all traditional Chinese medicine practitioners. Many had cross memberships, and could be divided into the following categories:

- trained in the Mainland;
- trained in Taiwan;
- trained in local schools of Chinese medicine;
- trained as apprentice of senior traditional Chinese medicine practitioners.

To have these members with such diverse training and political backgrounds agree on a single standard was like "building castles in the air".

With senior TCM practitioners Mui Ling-cheong (梅嶺昌), Tam Lin-kwan (談靈鈞), Kwan Chi-yee (關之義) and the Director of Health, Margaret Chan Fung Fu-chun (陳馮富珍) (Courtesy of *Sing Tao Daily*)

The Government understandably was adopting a non-chalant, non-interference attitude. As a British colonial Government, it knew that interference with Chinese customs was sacrosanct.

I personally was extremely passionate about giving TCMP a proper recognition. I believe that having proved its value and effectiveness for 5,000 years, TCM ought to be given a chance to develop properly, and to do good to our public. I also believe that those who trust their health and lives to TCM practise should be given adequate protection.

There were senior practitioners of TCM who were sitting on the Basic Law Consultative Committee who agreed with me, notably Mui Ling-cheong (梅嶺昌), Tam Lin-kwan (談靈鈞) and Kwan Chi-yee (關之義). But they were in a minority.

After being elected to the legislature, I arranged to have many meetings with the representatives of the 12 TCM organizations. They were all extremely courteous but they insisted that standardization for registration was unnecessary and impossible. They also questioned my motive – why would a Western doctor be so keen about registration for TCM practitioners? Did he have a hidden agenda?

As the saying goes, "crisis brings opportunities". In 1989, two members of the public died after ingesting the Chinese herb Gentiana (龍膽草) prescribed by a TCMP. This posed two serious questions – Was the TCMP who prescribed the "killing herb" properly trained? What protection an ordinary citizen had when he visited a "TCMP"?

This presented an opportunity for me to raise alarm to the public. I introduced an emergency debate in the Legislative Council to urge the Government to take action. Meanwhile as the Chairman of the Legislative Council Health Panel that oversees the Government's health policy, I formed a subcommittee to address the "standards" of TCMP. I persuaded two pro-China Legislative Council members, Hon. Tam Yiu-chung (譚耀宗) and Hon. Ho Sai-chu (何世柱) to co-chair this sub-committee, neither had association with the medical profession. I refrained from taking a leading role as I could easily be seen as having a conflict of interest.

An important first step was thus taken. Some years later an ordinance was enacted to register TCMP and regulate TCM.

The ordinance stipulated which TCMP could be registered in the future, how they need to maintain their registration status based on continuous professional development and certificate of good character. It also made provisions for currently practicing TCMP – those that had proof of active continuous practise for over 15 years would be given a full registration status, while those with 10 years of experience are enrolled practitioners requiring assessment after five years to qualify for full registration. No one, therefore, would lose their "rice bowls" because a new law is in force.

The Chinese Medicine Ordinance of Hong Kong was considered so comprehensive that the Singapore Government took it as a basis for their TCM registration.

The battle was won. TCM practitioners are now recognised health care professionals, and citizens who use their services are now protected. Today, the Government provides TCM services to the public medical care in 18 clinics, sometimes side by side with Western medical clinics. A stand-alone TCM hospital is now being planned.

How should TCM be further developed in Hong Kong? There are many who believe that TCM should be integrated with the Western medicine. I totally disagree. This will only tarnish the true value of TCM and demote it into a department of Western medicine. I believe that the philosophy and the principles of TCM practice should never be buried and laid to waste. Instead we should uncover the hidden scientific basis of TCM and dig deeper into this "hidden" treasure of our culture.

Chapter 7

Doctors & society

There were times, in the good old days, when the main aim of a doctor was just to save life. This is still true. Yet as medical services improve and human lives become even more precious and patients more demanding, the doctor's role gets more complex. Keeping someone alive is no longer good enough. Patients want to be able to return to their normal lives as soon as possible. Today, even that is not enough, patients want to return to total normalcy.

Let me give you some examples. An athlete is injured in a near-fatal traffic accident. His lower limbs are crushed. An amputation is done to save his life. But he needs to walk to do his daily chores, so he is provided with a pair of crutches. Then doctors want to return him to his profession. He is now given a modern artificial limb.

A recently married young woman is pregnant. Eight months into her pregnancy, she starts to bleed. It is diagnosed that she has extensive placenta previa. The bleeding is uncontrollable despite termination of her pregnancy. The only life-saving option was to do an emergency hysterectomy. She is saved. But she and her husband are devastated. They are desperate to have a baby for procreation and perhaps inheritance – the husband's father is a billionaire. Now they are in despair for without a womb, how is the wife going to carry a baby? Yet, as the saying goes, "where there is a will, there is a way". The sperms of the husband can fertilize the egg of the wife in a "test tube" and the embryo can "mature" in a "rented womb" of a donor. The child thus conceived, though carried by a surrogate mother, is genetically still the couple's baby. Yet life is full of unexpected outcomes. The surrogate mother develops a bond with the baby which she has carried for some nine months. She decides to keep the baby and refuses to hand it over to the genetic parents. That is the famous case of "Baby M", the pseudonym for the infant whose case was the first American court ruling on the validity of surrogacy.

Discussion on scientifically-assisted human reproduction

In 1986, I was appointed to chair a committee to study scientifically-assisted human reproduction (SAHR). Since medical science had greatly expanded the limit of what could be achieved, the committee was established to study the moral, legal, ethical and social aspects of SAHR, in particular the protection of the child conceived by SAHR.

The following cases give an idea of the many issues we faced and their complexity.

A couple has been married for five years. Despite their best efforts, including all sorts of superstitious practices, they remain childless. The parents-in-law are, as expected, anxious. They blame the wife, and threatened either to have the husband divorce her or to get him a concubine. The problem, however, actually lies with their son – he produces no sperms and is too proud to admit his "inadequacy". Finally, the couple decides on artificial insemination by a donor. This works and a son is born. Not long after that, however, the marriage turns sour. They can certainly get a divorce, but what protection will the baby have? He has been brought to the world not out of his own will, but to give pleasure and satisfaction to his parents when they still loved each other. He is in effect a "bastard" without a legal father.

In another case, a couple wants to have a baby. Yet the socialite mother doesn't want the trouble of getting pregnant and having to carry a baby. She has no intention to give up her work, albeit temporarily, for pregnancy. Nor is she prepared to forgo even a part of her exciting social life she enjoys so much. And the thought of having a less than perfect body after giving birth terrifies her. This is where the latest advances in medical science come in. An embryo can be formed by using the husband's sperms and her own egg, and then implanted in a surrogate mother. Well, there are medical risks and she needs to take medication to control the timing of her egg production but that is tolerable. What about the surrogate mother? Her needs must be adequately provided for but she resents any effort to control her. If the baby is born with congenital abnormalities, will the couple have a change of heart? If, on the contrary, the baby is perfectly normal, will the surrogate mother refuse to give it up? Remember the carrying mother is always the legal mother and any contracts the genetic parents sign with her before, during and after her pregnancy and delivery are non-enforceable. In such a context of complicated relationships, what protection will the baby have when he/she is being tossed around in a "no man's land"?

Things can be even more complicated. If the husband is "azoospermic", he has to "borrow" sperms from a donor. If the wife refuses to undergo the medical and surgical retrieval of eggs, the eggs have to be obtained from a "donor" and then implanted into a surrogate. In this case, "five adults" are involved. Who does the child conceived under such circumstances actually belong to? What is his origin?

In another instance, a lesbian couple wants to have a baby. One of them gets pregnant from donated "sperms". A child is born. When he gets older, he goes to school. But he is different from most of his classmates in one important aspect – he will always be "fatherless". This is of course not discrimination against same-sex marriage. Two adults exercising their rights to choose is not our concern. But what about their child, whose life is so radically different from the norm?

In view of all this, we put forward many recommendations, not on the scientific aspect of modern technology, but on the moral, social, ethical and legal issues, that this evolving society must seriously consider before moving ahead with the unstoppable progress it has made possible.

Some of these recommendations were:

- Assisted human reproductive technology should only be applied to married couples who can provide medical proof that it is not possible for them to have normal pregnancies;
- In the case of "artificial insemination by donor with the husband's consent" , the consenting husband is deemed the legal father of the child;
- Surrogacy is allowed for genetically-related embryos. There must not be any commercial dealings involved. Any contract between the commissioning couple and the surrogate mother is unenforceable;
- Sex selection by any means is prohibited
- The carrying mother is deemed the natural mother of the child born;
- The donor's identity must remain anonymous;
- Any child has the right to know whether around the time when he/she is conceived, whether the parents had been subjected to artificial scientific reproductive procedures;
- A statutory Reproduction Technology (RT) Council is to be set up to deal with details and regulate the procedure.

An RT Council was finally formed after the Artificial Reproduction Technology Ordinance was enacted and I was appointed its founding chair. I stayed on the whole movement including the Chairmanship of the Reproductive Technology Council for a total of 23 years.

These recommendations, I believe, were fair as they took into consideration the needs of the couple who yearn to have children and came up with measures to protect the child conceived under these circumstances.

Protection on industrial health

By the late 1980's and early 1990's, as Hong Kong's construction and light industries took off, it became obvious that workers were not provided with adequate protection and compensation for industrial diseases and accidents. In particular, there was little concern for occupational diseases such as pneumoconiosis and noise-induced deafness. It usually takes an extended period of occupational exposure for such diseases to develop and it takes some time before the debilitation surfaced. Therefore, though their effect on workers can be debilitating, it is often difficult to establish a correlation between the workers' disabilities and their work places.

Some old timers in Hong Kong still have a vivid memory of the iron ore mines in Ma On Shan in the 1950's and 1960's. Debris of the mining shaft, if inhaled, will in the long term lead to progressive fibrosis of the lining tissue of the lungs, resulting in loss of breathing reserves.

At that time in the late 1980's, the strong and solid infrastructures of buildings were often created by using caisson. Construction workers had to dig into the ground to produce a hollow cylindrical form some 20-30 feet deep until solid rock was reached. The work was usually performed by a husband and wife team. The husband started digging and lowered himself into the shaft, while the wife stood guard at the opening of the shaft with a rope to pull him up in case of danger. Meanwhile, she would use a pipe to suck up all the dirt. The inhalation of the dust contaminated air in an enclosed shaft damaged the lungs and resulted in "pneumoconiosis". In the days to come, the lungs of the patients would become fibrotic. Eventually they would lose all their respiratory reserves and become wheelchair-bound or bedridden, in need of continuous oxygen inhalation.

Basic protective gears were available. Workers were advised to wear proper masks. Regrettably, this advice was seldom followed, as wearing protective gear is uncomfortable in an enclosed shaft and it makes breathing more labourious.

Under such circumstances, as one of the medical professionals in the Legislative Council then, I launched a debate on the issue of industrial health, suggesting that the use of caissons be banned and workers be required by law to wear protective masks. I also proposed that they should receive proper compensation through a statutory pneumoconiosis compensation ordinance, with money to come from a levy on all construction building projects.

Workplace noise-induced deafness is another matter. Many light industries produce a lot of high-pitched streaking noises, extremely damaging to the hearing apparatus, and may, in the long run, lead to the loss of hearing. Again, the symptoms are not immediately obvious, but the long-term effects could be extremely debilitating. Basic protection was always available but unfortunately seldom adhered to. I therefore moved a motion to enact a law to enforce protection and provide compensation to workers for their hearing loss.

Perhaps it was because of this concern with workers' welfare that I was later appointed to chair a committee to look into the issue of working hours.

It may be said that I was over-stepping my bounds. Why would a surgeon and a Legislative Council member representing doctors and dentists take such a strong interest in public health? Well, I had always believed that as an elected member of the Legislative Council, I reprcsented not just my functional constituency but Hong Kong people, it would be my role to advise the Government on issues that I had some knowledge of and felt passionate about.

Proponents of universal suffrage and critics of functional constituencies will do well to remember that the expertise of the Legislative Council members representing their functional constituencies could be beneficial to the society at large. It is therefore the singer (the elected member) , not the song (the constituency) that matters.

Fighting AIDS

The first case of Acquired Immune Deficiency Syndrome (AIDS) was reported in USA in 1981. Then the disease spread like wildfire and threatened to become an epidemics terrible as any in recent memory. Worse, the virus had no preventive vaccine or drug cure. Only through education and precaution could the disease be contained. The Hong Kong Government was aware of the problem and acknowledged its seriousness. But she refrained from taking the lead in the fight against the disease. Instead, it enlisted the help of a non-government organization (NGO) to do the "pioneering work". With a subsidy of 15 million dollars from the Government, and another 15 million dollars donated by the then Royal Hong Kong Jockey Club, the Hong Kong AIDS Foundation was set up to promote the understanding of AIDS, and educate the public about its prevention. This was again a public health rather than a surgical issue, but as a medical practitioner, I felt compelled to take over the leadership of the Foundation from Mrs. Peggy Lam (林貝聿嘉), a former Chief Executive of the Hong Kong Family Planning Association and an active member of the Hong Kong Federation of Women.

The first education campaign we launched was to induce fear – "AIDS kills", hoping this alarming message would help prevention. Instead, it induced discrimination and inadvertently encouraged AIDS and HIV positive victims to go underground and shun "treatment". Fear of the disease among locals was sometimes so hysterical that some believed one could be infected just by shaking hands with AIDS victims or by sharing the same table for food or the same toilet seat.

I once took a dentist to lunch. He was a brave man and let it be known that he was HIV positive. But you could not say the same about the head waiter of the restaurant where we had lunch. He told me quietly that I would not be welcomed if I were to bring

Gathering with volunteers of the Hong Kong AIDS Foundation (Courtesy of Mr. Ducky Tse)

an HIV positive person as my guest again!

Our next approach was therefore "education". The purpose was to inform the public that HIV/AIDS could only spread through unprotected sex, sharing injection needles and mother to child transmission. This time, the campaign was also more focused, directing our resources at the workplace, schools and mothers as influential members of the family.

I believe all these efforts bore fruit. Hong Kong has a low incidence of HIV and an even lower incidence of AIDS. With the enactment of an anti-discrimination ordinance, I believe, HIV positive individuals need not go "underground". They can openly seek treatment in the Government hospitals at little or no cost.

As Hong Kong approached 1997 when it would be returned to Chinese sovereignty, it was obvious that communication and contact with the motherland be it business, academic or pleasure-related, would become more and more frequent. This raised concern for "any wind that blows from the north will send chills down our spine".

A case in point was China's policy towards HIV/AIDS. Initially, the official stand was one of denial – AIDS is a foreigner's problem. Little did officials know the disease was spreading rapidly – through drugs and the sharing of needles in places like Xinjiang, Yunnan and Guangxi, and via sexual contact and mother to child transmissions. In Henan, there was an outbreak that spread from the sale of blood.

Hong Kong was naturally concerned. While we had a low infection rate, if the situation across the border got out of hand, Hong Kong, with its proximity to the Mainland and hundreds of thousands of people crossing the border for business or pleasure every day, would not be spared.

With our experience and success in launching education and prevention campaigns, the Hong Kong AIDS Foundation believed that it was duty-bound to help the Mainland understand the disease better, and in all its complexity.

We took two important steps. Firstly, introduce the international AIDS community to China, or rather introduce China to the international AIDS community. Secondly, use our very limited manpower and resources to assist China to come up with policy directions for the effective education and prevention of HIV/AIDS.

Dr. Shen Jie sharing her experience on AIDS in China at a Vancouver conference

In 1996, at an international AIDS conference in Vancouver, we organized a satellite symposium on "AIDS in the Chinese Community". Speaker from AIDS organizations for Chinese in Australia, Malaysia, Singapore, Taiwan and, of course, Hong Kong talked about their AIDS problems and how they fought the disease. For the first time, officials from China spoke out on the problem. The Secretary General of Chinese Association of STD (sexually-transmitted diseases) /AIDS Prevention and Control Dr. Shen Jie gave an official address on HIV/AIDS in China. The taboo was broken, and China woke up to the problems of HIV/AIDS.

The subsequent work that the Hong Kong AIDS Foundation did in China was excruciatingly difficult but ultimately successful and rewarding. We used our limited manpower and financial resources to launch a scheme to "train the trainers". We provided training and financial support to designated "trainees" chosen by the Chinese Association of STD/AIDS Prevention and Control to come to Hong Kong to work side by side with our staff for two weeks. On their return to their place of origins, they could use the knowledge and techniques of education and prevention they learned in Hong Kong to train more people. As of today, we have directly and indirectly trained over 20,000 compatriots. We stay in touch with them through an alumni association which meets once every two years in different cities in China. We also cooperated with Party schools and signed an official Memorandum of Understanding (MOU) with them. Ms O.C. Lin, (連愛珠) the Foundation's Chief Executive then, deserved credit for doing most of the organization work and arrangements.

Little by little progress was made. On December 1, 1997 after celebrating the World AIDS Day in the bitter cold on the steps of the Great Hall of the People, I accompanied China's then Minister of Public Health, Chen Minzhang (陳敏章) to visit an AIDS patient in a Beijing Hospital. This marked the first time for Chinese officials to embark on such a mission. He shook hands with the patient in the full view of local media and TV networks, representing an important ground-breaking step towards dispelling fear and discrimination against the disease in the country. With the help of the international renowned HIV/AIDS expert Dr. David Ho (何大一), we managed to reduce the incidence of mother to child transmission through early administering of anti-retroviral medication to the mother (cocktail drug treatment) and promote the treatment to

With Chen Minzhang celebrating the World AIDS Day in Beijing

With Dr. David Ho and Aaron Kwok at an event of the Hong Kong AIDS Foundation

HIV positive patients in China.

Today, China is making an all-out effort to educate her people on the prevention of HIV/AIDS. The First Lady Peng Liyuan (彭麗媛) is now actively involved as the Ambassador of AIDS in China.

In no way am I glorifying the Hong Kong AIDS Foundation. We believe, as an NGO in Hong Kong and therefore, part of China, we have an obligation to share our experience and expertise gained through an earlier start with our compatriots. With our help, China had finally come to grips with the urgency of AIDS and the extent of the problem. We also believe that only through understanding and acceptance can prevention and education have a chance to succeed in controlling the imminent epidemic. It benefits Hong Kong too as its number of HIV/AIDS diagnoses has been somewhat under control.

Chapter 8

We are all getting old

"Ageing is a destiny", Alan Greenspan once said. It is true! You cannot stop ageing. And if Hong Kong is ageing more rapidly than the rest of the world, it's because we live longer than everyone else.

Our average life expectancy for the female (2017) is about 86 and for the "stronger sexy" is about 82. One in seven in Hong Kong today (2017) is 65 and above and in 2030 the percentage will be one in four. This may come as a surprise to many. Hong Kong is a city of fast pace and high stress. It is extremely crowded with people and the air quality is far from satisfactory.

Perhaps the longevity of Hong Kong people can be attributed to its advanced health care and superior medical services. Our infant mortality rate is 1.5 per 1,000 registered live births and our maternal mortality rate is 1.6 per 100,000 registered live births. But ironically, an ageing population poses many challenges to society, and vindicates the prediction that "the success of the 20th century contributes to the problems of the 21st".

As always, challenges come with opportunities. Yet the society and the Government often take on a doomsday attitude. We are told repeatedly that our "dependency ratio" is rising, meaning there will be more and more senior citizens who are "non-income earners" to be supported by fewer and fewer young people – who are income earners.

I don't challenge these figures, but it is most unfair to characterize those over 65 as "dependents". The society needs to think out of the box to tap into the expertise, experience and productivity of its members aged 65 and above. In fact, it is entirely foreseeable that many senior executives of major corporations who are now, say, 55, will still be taking the helm in ten years.

The idea of "active ageing"

The truth is the elderly of today are healthier, more active, sociable and financially independent than ever. To provide for these people by handing them welfare is therefore not only erroneous but discriminatory. Yes, they need care and concern, but certainly not in the form of welfare that hurts their self-esteem.

When I was asked to chair the Elderly Commission which advised the Government on the formulation of a comprehensive policy for the elderly, I made a few visits to elderly day care centres. The experience wasn't entirely pleasant. Most of those who stayed there were old ladies who spent their time watching old Cantonese movies on television and folding paper flowers. They were being taken good care, I do not deny, complete with hot meals and dim sums. But where are the senior males?

Talking to senior citizens at the Pilot Neighbourhood Active Ageing Project

My heart often sank when I walked through the side streets and open spaces of Wan Chai, Yau Ma Tei and Western district, and saw the old men there. They seemed to have nothing better to do than idling around, scanning through newspapers or magazines that were at least 2-3 weeks old or playing a game of chess with one another.

I asked them how they would like to while away their time.

Their answers surprised me. No, they did not want welfare. What they wanted was to learn something new, so that they would be equipped with modern knowledge and stay in touch with the society and their grandchildren. Learning English and how to use the computer and smartphone were their top priorities.

That was when I got the idea of "active ageing".

Hong Kong has the highest elderly institutionalization rate, i.e. a high percentage of elderly are placed in elderly homes providing different levels of care. This is to be expected: our living environment is not exactly conducive to caring the elderly at home; young members of the family have to work; community services, though available, are haphazard. Furthermore, if an elderly is institutionalized, he/she can apply for Comprehensive Social Security Assistance (CSSA) to cover the cost, and their children who support them get a high rate of tax deduction.

Yet, ask any old man or woman and they will tell you they want to spend their twilight years staying with their families at home, and in a community and neighbourhood they have come to know well.

Ageing "in place", therefore, should be our goal.

Something must be done to ensure that "ageing in place" be given top priority.

There are in general two types of elderly institutions. The publicly-funded ones (subsidized and contract homes) have more superior facilities, yet their places are limited. After all, it is neither possible nor desirable for the Government to provide facilities for all her senior members. Most of the elderly institutions are privately-run and they differ greatly in their quality of service, notwithstanding the Government's requirement for them to comply with a basic standard of service. Over 80% of the inmates in these private institutions are on CSSA. That means these institutes can only afford to provide the basic needs of their inmates. Ironically, under the strict regulations of CSSA, any extra financial support to the recipient, say from family members, would lead to a reduction in the amount of assistance provided by the Government.

Could something be done to bypass the regulations? How about calling it something else, *not* CSSA?

Another problem that makes "ageing in place' so difficult is that while one third of Hong Kong people live in public housing, many of these estates have no elderly home, not even in their vicinity. The elderly who need institutional care are sometimes sent to elderly care centres far away from their homes and communities. For example, an old man who lives in Chai Wan is admitted to an elderly home in Kwun Tong.

Discrimination against the elderly does exist in Hong Kong. In 2003, as the Chairperson of the Elderly Commission, I was approached by a group of residents from a public housing estate in Kwai Tsing district led by their elected member of the Legislative Council. They were opposed to allowing the podium of their estate to be used as an elderly home. The podium of their housing estate had been left unrented or unused for some time. It could be a perfect site for a new elderly home. A proposal to turn this idea into reality had been put forward by a consortium experienced in providing elderly services. The application and building plan were approved and construction was underway, until it was met with fierce opposition from the local residents.

They said old people were noisy, dirty and slow. They were

also accident-prone and could easily be knocked down by the energetic young people in the neighbourhood. How farcical! The project was finally abandoned.

The Elderly Commission therefore set four policy directions as its goal:

- Promote active ageing;
- Promote ageing at home or in community;
- Change CSSA for the elderly to "money that follows the elderly";
- Persuade the Government to designate areas in public housing estates for the construction of elderly facilities.

The Elderly Academy

Schools are always the best place to learn. They have the facilities, the software and the hardware. Most schools finish their classes in the early afternoon. Their classrooms are therefore available in the late afternoons, Saturday afternoons and Sundays.

The availability of classrooms and the enthusiasm of the elderly for learning are a perfect match. The fact is that many senior citizens did not have a chance to go to school when they were young. For them, learning new knowledge in a school environment would be a "dream come true".

But who would be the teachers? Well, why not let the school' students play the role of "little teachers", sharing what they know about computers, fad words and the English language? And this would be a great way to promote inter-generation harmony!

Graduation ceremony of the Elderly Academy

This was how the idea of the Elderly Academy was conceived as a means to put the concept of "Active Ageing" into practice. Participating primary and secondary schools would make their facilities available after school hours, and their students would become "teachers". As for the "students", they were supplied by established reputable elderly care centres. The scheme was given a shot in the arm by the Government with an injection of funds approved by the Financial Secretary.

Today, we have over 130 such academies throughout the SAR. Along the way, tertiary education institutions joined in to provide two types of programmes – a full time degree programme(Lingnan University) and a certificate programme. Senior citizens can enroll in either existing university courses (City University of Hong Kong, Open Univeristy of Hong Kong and Hong Kong Shu

Yan University) or programmes tailor-made for them (University of Hong Kong, Hong Kong Polytechnic University) . Apart from the opportunity to learn, they are often provided with "student cards" and given access to the universities' canteens and libraries. Some may even have the chance to lodge for 1-2 nights in a university hostel. It is a bonanza for those who have never studied in a university. Now they find out for themselves what a university life is all about – attending classes, tutorials, visiting libraries, eating in student's canteens, staying in student hostels and taking not just pleasure but justified pride in gowning themselves at a "graduation" ceremony attended by their spouses, children and grandchildren. They are not the only ones who benefit. The "genuine" university students who become their "buddies" and guides learn from their life experiences.

Different academies organized different functions for their senior students – graduation ceremonies, sports meetings, high table dinners, painting exhibitions – life begins again!

This is very different from the U3A (University of the 3rd Age) in the US and Australia. Every U3A is a conglomeration of retirees – "students" are retirees, "teachers" are retired university lecturers, and even the staff are retired university administrators. The Elderly Academy, however, is all about inter-generation harmony. At a time when society is getting more and more divisive, a better understanding between the older and younger generations is necessary not only for social harmony but political consensus.

That's what I call "active ageing" (Courtesy of Mr. Ducky Tse)

The concept of "Active Ageing" is given a big push by the Government's policy to provide concessionary rates to senior citizens when they travel in Hong Kong. For two dollars per trip, the elderly can take the MTR, public light buses, regular bus lines – a policy that really deserved to be applauded. With cheap travel, the elderly are encouraged to move around, visit their grandchildren and enjoy social gatherings with peers. Active ageing is finally set to take off!

We look forward to all the schools, privately and government-funded, to join the Elderly Academy. We look forward to extra funding from the Government or charitable organizations like the Hong Kong Jockey Club to set up a central Elderly Academy coordinating centre to better coordinate resources, so that there will be an Elderly Academy in every neighbourhood for all senior citizens.

The innovative way we deal with ageing hasn't gone unnoticed. It has been a model for the Singapore Government, the Thailand Education Minister as well as many elderly welfare bodies in Taiwan.

The world has witnessed many changes, and throughout the world societies are ageing. There is also a change in the needs and culture of the young people. They are now more demanding and more self-centered. It is a "me" and "I" society rather than a "we" and "us" concept. In the past, young people addressed the elderly by their seniority in the family – grandpa, grandma, father and mother. Today, in the US in particular, many address their seniors by their names. In bygone days, the elderly were authority figures who wouldn't think twice about asserting their authority. The young family members addressed all senior members every morning and at the dining table. Not now anymore. Today, the elderly have to learn to treat the junior family members and young people in general as their peers or friends. The time-honoured values of filial piety seem increasingly a relic of the past.

In Hong Kong, the low birth rate (less than one child per couple) , the shrinking size of families, the cost and hardship of raising a child (many opt to have a pet dog than a child) , and the nano-sized living quarters all contribute to an increasingly popular social phenomenon of senior citizens living alone, many of them end up as the so-called "hidden elderly". They live alone and tend to withdraw from the society, locking themselves up and burying themselves in old newspapers, dirt and their meager belongings.

It is a disgrace to a relative affluent society. They will not even answer the door when welfare officers knock. Attempts to gain entry by social worker of NGOs are met with the same fate – these social workers are usually too young to be able to build up any rapport with the elderly. We therefore decided to tap into the neighbourhood spirit and community network. We asked the more active senior citizens living in the same housing estates to knock at their doors, and engaged them in small talks and help them to buy daily necessities. After repeated attempts, they often succeeded in gaining the confidence of the "hidden elerly" and become their friends. Some of them even came out of their isolation and joined "kai fong" activities.

To have a better understanding of these elderly people, we commissioned the University of Hong Kong to conduct a study on why, despite the prevalence of "ageing at home", there is still a high rate of admission to elderly institutions, and to make recommendations on how to address the problem.

In short, the study recommended that the Government should subsidize the elderly living at home with Home Care Service Voucher for the elderly. The Government was interested. A pilot scheme was launched in a few districts but the results were not satisfactory. I believed the problem was two-fold. Most senior citizens had little idea of what services they needed and even less idea of where they could get these services. To address this problem, two things need to be done. NGOs specialized in elderly

High-table dinner at HKU's Loke Yew Hall

services should help the elderly identify what they need and to "purchase". Elderly homes should expand their services and reach out to those who stayed at their own homes. It is time to think out of the box to ensure that "ageing at home" becomes reality.

For those who have to work for most of their adult lives, the thought of retirement is very enticing. You no longer need to stick to a military-like discipline everyday – get up at 7am, dress properly, rush out for a quick breakfast at McDonald's or "cha chaan teng" and squeeze yourself into the MTR. You can leave town and visit places you yearn to see, and do things that you have always wanted to do but have no time.

But retirement life isn't always what it is cracked up to be. You are in your pajamas up to 11am and your wife is not pleased. The breakfast she prepares is not exactly your cup of tea. You want to go on holiday trips, but how many trips can you take and with whom? Most of your friends still have to work and, holiday trips are expensive. You have some savings, but no regular income. A dollar spent means a dollar less saved. You become a miser. You only buy those items that you really need and in small doses. Such financial pressure can lead to psychological depression and some funny behaviours. For instance, before retirement, you go to the supermarket with the sole purpose of buying toilet papers but leave with a trolley full of unnecessary items. Today, after retirement, you spend hours touring the supermarket in search for the cheapest roll of toilet paper. Being able to purchase freely, though not overspending, is sometimes an efficient way to relieve stress and depression. It makes you "feel good" about yourself.

You used to believe that the 24 hours of the day were not enough for you to get your work done. Now 24 hours is too long. You feel unproductive and out of touch. Your children have to take time off their busy schedules to see you. It used to be the other way round. You feel you are not needed both by your family, your friends and the society. You consider yourself a "burden". How sad! Do you still want to retire?

According to the Government's population policy, Hong Kong should have more young people to provide for the needs of the increasing number of senior citizens. But a well thought-out population policy should not look at old people as a burden. Instead, it should propose measures to make good use of their experience, expertise and productivity.

Hong Kong does not have a legal retirement age. However, civil servants have to retire at the age of 60, and those in the

Sports day for the elderly

discipline forces, at the age of 55. This is understandable, for back in the colonial days when civil servants were sent from the United Kingdom to work here, they wanted to retire early to go back to their own hometown sooner rather than later. Furthermore, in the 1940's and 1950's, not many lived to the ripe old age of 70.

It is very different today. People in their 60s are actually in the prime of their lives. They have accumulated valuable experience and many no longer have to take care of their families. That makes them a formidable force to lead and to train the young guys in the offices before they are ready to take over.

The manpower shortage of public hospitals tells a sad story of missed opportunities. It has become customary for the Government to complain of the lack of manpower at public hospitals. Yet when a consultant doctor or a senior nurse reaches the age of 60, they are given the marching order to retire, along with the precious experience they have gained over the years. Meanwhile, the management of hospitals is all out to recruit young medical workers with little experience and knowledge of the needs of Hong Kong patients. How farcical!

I have always called for flexible retirement. If someone wants to retire from work for health or other reasons, or someone to get another job, he or she should be given the freedom to do so. But if they are fit and good at their jobs, why force them to retire?

It may be argued that if people do not retire, others will not be able to move up the "ladder". This is *not* true. Senior managers at the point of retirement can always move aside and take up a purely professional role to help the up and coming move up the career ladder.

My experience in the Elderly Commission and later as chairman of the Elderly Academy Foundation Fund (which I founded) gave me great pleasure and satisfaction. Together with Tam Yiu-chung (譚耀宗) (the Founding Chairman), Prof. Alfred Chan (陳章明) (my successor who now chairs the Equal Opportunities Commission) and then Dr. C.C. Lam (林正財) (now a member of the Executive Council in Mrs. Carrie Lam's (林鄭月娥) cabinet), I revolutionized the concept of elderly care. It should no more be "welfare". By promoting the concept of "Active Ageing", we demonstrate what it means to be a senior citizen in the 21st century and why they are not a burden to their families and society as they still have a lot to contribute with their dignity and confidence undiminished by old age. Caring for the elderly, therefore, is not a welfare.

The elderly are often misunderstood. They are seldom

demanding, and many are actually proud. They shun welfare. I remember a visit to an old couple who live on the top 7th floor of an old building in Kowloon City. There is no elevator. The husband in his late 70s has not totally recovered from a stroke. The old lady in her early 70s, walked up and down five or six times a day to buy food, gather cardboard boxes and sell them for money. It was a difficult climb for me. When I finally reached the 7th floor and had regained my second wind, I offered to make arrangements to put them on Comprehensive Social Security Assistance (CSSA). I was turned down flatly. "With the old age allowance and earnings from the sales of cardboard boxes, we could manage adequately", she told me with a proud smile. She also told me, to her great delight, a cook from a nearby Chinese restaurant would bring some leftover soup to them two to three times a week.

The elderly are easily satisfied. Just sit down with them, holding their hands, show your care and concern. This is enough to brighten their days.

Chapter 9

Repaying my alma mater

The University of Hong Kong (HKU) is one of the major icons of Hong Kong. The first and for a long time the only tertiary education institution of the city, it was an institution that was endorsed by both the then Chinese and British Governments. It has remained unperturbed by tumultuous social changes and political turmoil for over a century, training students – not only elites but also sons and daughters of grassroots workers and ordinary people – for "Hong Kong, China and the world" to take up leading roles in academia, business, finance, professional and the Government.

The University is a legal entity governed by the Hong Kong University Ordinance which was enacted in 1911. On March 30, 1910 the foundation stone was laid by Sir Frederick Lugard, the then Governor. In 1912, the University admitted its first student.

It is therefore of no surprise that the celebration of its centenary became a major event that spanned over three years. The highlight was the centenary gala dinner on December 18, 2011 held on both the old and new wings of the Hong Kong Convention and Exhibition Centre. Over 5,000 guests attended, including alumni from round the world, senior government

At the HKU Centenary Gala dinner (Courtesy of the University of Hong Kong)

officials, major benefactors, dignitaries and representatives from sister institutions. Most were related to the HKU in one way or another, and they were all proud of their connections. The HKU's spirit was on remarkable display.

For me, it was the proudest moment as the Chairman of the Council, HKU's highest governing body, and it gave me great pleasure to be at the helm of this majestic institution on her 100th anniversary. I had finally repaid my *alma mater*!

You can't blame a 100-year-old institution for resting on its laurels. HKU had its share of challenges but had always emerged from them stronger. Yet the world is not what it used to be and people demand transparency and accountability. They want to have a say in the running of the institution. After all, the University of Hong Kong is publicly funded and has at least five groups of stakeholders – the governing body, the management, the alumni, the students and the public. What these stakeholders share is a strong belief that the University's institutional autonomy is sacrosanct.

In the late 1990's, I was asked by several staff members of HKU, some then Council members, and some alumni, to join the Council. I saw this as an opportunity to serve the place that had nurtured me. I stayed in the Council for five years before I became its Chairman in 2010.

HKU is one of the eight tertiary education institutions funded by the University Grants Committee. Each of these institutions is governed by a different ordinance and is different from one another in its structure and organizational hierarchy. The head of HKU is the Chancellor. In the colonial days, the Governor was invariably its Chancellor. This tradition continues after 1997 and the Chief Executive becomes HKU's Chancellor. The position is by no means ceremonial. It carries with it genuine executive power. A typical example is the award of honourary degrees – the highest honour that a university can bestow. At HKU, the recommendations, made by an honourary degree committee, are submitted to the Chancellor, for approval or veto as he deems fit.

The basic structure of governance of HKU consists of the Court, the Council and the Senate. The management is headed by the Vice Chancellor, now also called the President, and the senior management team (SMT) .

The Court is basically an advisory body, headed by a Pro-Chancellor appointed by the Chancellor. It consists of members of the Council, the Senate, elected members from the Legislative Council, elected members from the alumni association and local dignitaries as appointed members.

The Council is the supreme governing body. It is headed by the Chairman appointed by the Chancellor and assisted by six appointed members from the Chancellor and six appointed members of the Council, elected representatives from the Legislative Council, postgraduate and undergraduate students, elected teachers and non-teaching staff members. (The composition of the Council follows the recommendations set out in the Fit for Purpose Report 2003 submitted by a panel consisting of three members appointed by the University – This panel was set up by the University Council on July 30, 2002 to conduct an overall review of its governance and management structure. The panel consists of external members who are familiar with the education system in North America and the Commonwealth and local members who understand the local environment and situation in Hong Kong. The members of the panel included Professor John Niland, formerly Vice Chancellor and president of the University of New South Wales, Professor Neil Rudenstine, formerly Vice Chancellor of Harvard University, and the Honorary Chief Justice Andrew Li (李國能). Every member of the Council sits there as trustees. They do not need to represent the constituency that elected them, yet they all want the best for the University. The Council formulates the management policy, allocates resources, approves academic recommendations of the Senate and investigates complaints of staff and matters of discipline through the Grievance Panel. It also selects and appoints the Vice Chancellor. The Council is essential in the highest governing body of the University.

The Senate is chaired by the Vice Chancellor and makes recommendation on all academic matters for the approval of the Council.

For a whole century, this oldest bastion of high learning stayed in her comfort zone and operated by the principles set out in its statutes of governance. Rules and guidelines were there, but for some reasons, few were implemented. Basking in the glory of being Hong Kong's highest learning institution, HKU saw little need to adapt to the rapid changes of society.

On joining the Council of this great institution, I was amazed. Meetings of this supreme governing body were haphazard. Yes, there were scheduled regular meetings. Many, however, were cancelled at the last minute because "there was no agenda". Council members knew very little about the University. When I casually asked some departing Council members how many faculties there were in the University, many said they did not know. When I conversed with senior staff members and asked how well they knew the Council members, most could not identify them or didn't even recall their names. This was a pity. As trustees, Council members have a duty to take ownership of the University, yet how could they do so if they do not have any idea of what is going on? Council members come from diverse backgrounds but they are all leaders in their own fields. They can be invaluable assets if they have a rapport with the staff.

The University has grown many folds since the days of Dr. Sun Yat-sen (one of the first graduates of the Faculty of Medicine) – from two faculties (Medicine and Engineering) to ten and numerous departments and schools, from a single main building housing the Loke Yew Hall to estate and campus extending all the way to Sassoon Road in Pokfulam.

Then there are the people. In the year 2016/2017, there were 16,809 undergraduate students, 11,935 postgraduate students, 7,786 academic staff members, and 3,901 non-academic staff members. There are more than 200,000 alumni. Each group

With former Governor Lord David Wilson (Courtesy of the University of Hong Kong)

claims ownership of the University. They demand to know what is happening to their beloved institution. They want a say in policy-making and the running of the University. Most importantly, they want their ideas and suggestions to be taken on board.

All this falls within the ambit of one person – the Vice Chancellor.

Most will agree that a University should be academically-oriented. In this respect, the Vice Chancellor is assisted by the Provost and the Pro-Vice Chancellors who look after research and other academic affairs supported by the Deans (Heads) of the various faculties. Yet a big institution like HKU has much more to consider than academic development – campus security, fundraising, finance, estate, campus development and corporate affairs. Streams of dignitaries and heads of states often pay visit to us, sometimes to be conferred honourary degrees. Special protection and security measures must be taken to ensure the guests' safety and the reputation of the University. To do this effectively, someone needs to be put in change.

All these presented new challenges to the University. Reforms must be made to meet with the expectations of the public.

In early 2007, as Chairman of the Committee of Enquiries of the Council, I was referred by the Vice Chancellor to a complaint lodged by a private patient concerning irregularities in charging by a senior member of the Medical Faculty. There was obvious *prima facie* evidence and the matter was therefore referred to a law-enforcing body. The case was finally brought to the attention of the district court and that staff member of the Medical Faculty involved was convicted on charge of "Misconduct in Public Office" (DCCC 440/2008). That staff member, after performing a service to the patient, asked the patient to pay fees into an account of which he was the sole holder. And as the sole authorized signatory, he used letterheads of HKU and Queen Mary Hospital for receipts while never informing the University of these payments.

While outside practice by faculty members is allowed with approval, this case exposed the inadequacies of existing processes and procedures concerning billing arrangements for charging private patients. And while there were regulations on the declaration of interest, they had never been strictly enforced. In particular, the following problems were highlighted:

- Clinical records were not kept properly and therefore no proper billing arrangements could be made. As a result, patients could be asked by individual staff of the University to pay into an account not belonging to HKU and the Hospital Authority;

- There was no procedure to regularly check whether there were discrepancies between the billing records and the medical records;
- Declaration of conflict of interest policy was never implemented.

"Outside practice" has always been a bone of contention. It may be argued that members of the University are obligated to provide a service which is deficient in the market. In the case of a medical doctor or dentist, it could be for a humanitarian purpose. Sometimes the practice is to assist another surgeon to complete an extremely difficult surgical procedure. Yet it may also be argued that University's full time employees receive full salary from the public coffers. His expertise therefore should mainly benefit the public. Serving and charging private patients means that he is obtaining double benefits at the expense of the public.

The Grievance Panel therefore recommended that the management should provide and implement clear guidelines and regulations on declaration of interest, application and approval for outside practices, apportion of income between the Department, the Faculty and the University and the final disbursement to the relevant staff members and how the money should be used.

It was hoped that these recommendations, if accepted, would close the loopholes. Yet rules and procedures alone are not enough. It is the integrity of people that serves as the best safeguard against abuse.

Once I took up the Chairmanship of the Council, I instigated a programme of reforms. I opened up Council Meetings to the senior management – the SMT and the Deans (hitherto only the Vice Chancellor attended as a member of Council). The purpose was to give senior management a better understanding of Council members' views on various issues and how they feel about the University. I also wanted the senior management to know the Council members better so that it could make better use of their experience and expertise. I therefore made arrangements for the two sides to meet regularly, often on social occasions where they could get to know one another in a more relaxed atmosphere. Not everyone was thrilled. Some regarded my leadership style as too hands-on. But I was undaunted, and I insisted that the Council should meet every month on a regular basis.

A university in the society is like a microcosm in a macrocosm. Changing values and attitude of its stakeholders reflect similar changes in the society. The year 1997 was a watershed. Before that, we in Hong Kong had no say in how Hong Kong could be run. We were, as they say, living in a borrowed time and a borrowed place. 1997 should have given us a new lease of life. We are now supposed to be the master of our own house and should be able to determine our own destiny under "One country, two systems". But things turn out differently. There are people in Hong Kong who have absolutely no trust in the Central Chinese Government and a total lack of confidence in the HKSAR Government. They develop a culture not of compromise but confrontation. What's more, in their vocabulary there is no "us" and "we", only "I" & "me".

Perhaps a few incidents that occurred in HKU during my term as Chairman of Council could lend credence to these statements. They also showed how oblivious the University was to the changes which had our society increasingly polarized and confrontational.

Centenary Ceremony

Little consideration was given to risk management. When major functions were held, individual departments worked on their own without much communication with one another. There was no central coordination – nobody was in command!

In the summer of 2011, we learnt that a senior leader of the Central People's Government would be coming to Hong Kong to officiate at the opening of the new SAR Government headquarters and might also pay a visit to HKU. He would also bring good tiding. The senior management of the University was jubilant and rushed to plan an impressive welcome. It was finally decided that it should be a congregation type of ceremony with all participants in full academic regalia. And it should also be a celebration event of the University's Centenary. The guest of honour would address the gathering, to be followed by a response from a senior representative of an overseas university. The aim was to showcase to our guest of honour a proper Western-style procedure protocol of a solemn academic congregation. The guests invited to attend would include the University's major donors, academics, and non-academic staff. Students would not be left out, and student's representatives were chosen to speak to the guest of honour. The senior member of the Central People's Government was confirmed to be Li Keqiang (李克強), Vice-Premier of the State Council.

On the big day (August 18, 2011), everything went without a hitch. It was considered a "well organized" and "impressive" ceremony. Yet behind the scene, the view was quite different. The following was what went wrong and, with hindsight, what should not have been done or happened:

- A congregation-style ceremony should never have been arranged;
- In a proper British style ceremony (to which HKU is accustomed), the host (in this case, the Chancellor) invariably sits in the middle with the guest of honour on his/her right. The arrangement in Mainland China is, however, different. The guest of honour sits in the middle. In this congregation ceremony, the central seat was occupied by the guest of honour and the host (the Chancellor) on his right;
- Since the guest of honour is the Head of State, it is just natural that he would be given a high level of security. Responsibility was taken over by the police. Yet this was regarded by some as infringement of the University's institutional autonomy. And it's true that the University usually takes care of its security, and police would come only upon request;

Li Keqiang's visit to HKU (Courtesy of the University of Hong Kong)

- The guests who were major donors were all seated in the front row;
- While it was the intention of the students to stage a demonstration and hand a petition letter to the guest of honour, the designated demonstration site was far away from the approach route of the guest of honour.

These arrangements, reasonable as they were, left the University wide open to attacks and accusations:

- We gave up our institution autonomy by giving in to the Liaison Office of the Central People's Government;
- By allowing the police to come into the campus to take over security, we, again, surrendered our institutional autonomy;
- The seating arrangement implied that we are kowtowing to the rich and famous.

These accusations, though unfair, did mean something. First, they reflected a total lack of trust in the Central People's Government. Second, the anti-rich sentiment, fueled by the widening gap between the poor and the rich in Hong Kong, was on the rise. Third, there was growing concern with the inflation of police power. These were the genuine issues and concerns. Protecting HKU's institutional autonomy was just the convenient excuse and red herring.

The students (both undergraduates and post-graduates) staged a rally. Hundreds of them attended, shouted slogans and one by one they made speeches ridiculing the University's arrangements and decisions. They requested the Vice Chancellor and the Chairman of the Council to attend to answer queries as a means to vent their frustrations, and that was what we did.

As the Council chairman, I called for the setting up of a Review Panel to look into these issues and to make recommendations. The membership of the Panel included representatives of all stakeholders. It was hoped that it would be chaired by a senior judge, but no judge wanted this hot potato. The undergraduate students declined to participate too. They were encouraged to give their views – they never did.

The Panel's work was in no way easy. Its assessment and recommendations must be sensible and politically informed, and take into consideration the University's century-old culture. The final report found all the allegations groundless and unsubstantiated and recommended that the University should bring its operations and management in line with modern practices and expectations.

I must express my heartfelt gratitude to the committee, in particular to the chairperson Dr. Lester Huang (黃嘉純), a devoted member of the Council. His attention to details, befitting his legal background and impartiality were admirable. It also demonstrated how hard the Council members, as trustees of the University, worked for the best interests of the institution.

The Review Panel (on the Centenary Ceremony held on August 18, 2011) made many useful recommendations in its report. There was one, however, that struck a chord in my heart. There is a need to have someone to look after non-academic issues of the University. Accordingly, the position of a Vice President, a Pro-Vice Chancellor for administrative affairs was created by the Council and before long a qualified person was found to fill the post. I felt vindicated as I'd been arguing for the necessity of this position for some time. Nothing clarifies the mind like a crisis.

To me the whole saga was a "storm in a tea cup". Yet it brought home the fact that time had indeed changed. Today, nothing can be taken for granted and, despite the best of intentions, things can easily be politicized.

A University is definitely not just a place for high learning. It should also be a cradle for free thinking and free expression. University students are young and idealistic. They are understandably passionate about their beliefs and have strong opinions on the society. They want their views heard and taken on board. Yet any actions they take to promote their ideology must comply with the basic principles governing behaviour in a civilized society. The adult society should provide them with guidance and urge them to act with restraint, instead of adding fuel to their fire and leading them astray.

Successor of the Vice Chancellor

When Prof. Tsui Lap-chee (徐立之), the Vice Chancellor, finally decided not to take up a third term in spite of my repeated attempts to persuade him to do the otherwise, he sent me a note to the effect just prior to a Council meeting. I immediately knew we had another difficult task at hand.

Speculation about Prof. Tsui's decision abound. This was inevitable because it was made not long after Premier Li's visit and the controversy that followed. As expected, speculations soon spiraled into rumours. "The Vice Chancellor cannot stand the heat of the kitchen"; "The Vice Chancellor is forced to leave by the Liaison Office of the Central People's Government." These speculations were self-perpetuating. They refused to die down even after the Vice Chancellor openly expressed that he had done two terms (each for five years) and it was about time he returned to his basic scientific researches. Prof. Tsui showed great commitment to the University by staying on after his term expired until his successor assumed duty. To honour Prof. Tsui Lap-chee, I proposed and the Council unanimously supported that one of the four new residential colleges in Kennedy Town be named "Lap Chee College".

But the drawn-out saga of looking for the next Vice

Chancellor had just begun.

One would have thought that as this would be the 15th Vice Chancellor since HKU was established, there should have been some established criteria and selection procedures in place. No such luck. A thorough examination of the HKU Ordinance and the HKU statutory book gave no clue. The only mention of Vice Chancellor appointments appeared in our statute: "The Vice Chancellor shall be appointed by the Council after consultation with the Senate".

Obviously, something needed to be done. Again, the burden fell on the Council. Given how politically-charged the environment was, every step and action taken in the selection process must be transparent and all stakeholders must be given the opportunity to give their input.

We took the followings steps:

Step I: Begin consultation to determine the type of Vice Chancellor HKU needs. The future Vice Chancellor must possess the following areas of excellence: a strong and universally-acknowledged academic background; experience in management; a good knowledge of and an experience in fundraising and an ability to understand the political needs of the society.

Step II: The formation of a "Search Committee" to seek and collect recommendations on candidates both locally and internationally. A headhunter firm will be engaged to help. This Committee will sift through all possible candidates and assess them in terms of their academic backgrounds. It may even have to interview the candidates in their overseas working environment. This important task was given to senior professors of the University who were themselves selected by professors to preempt the accusation of working in a black box. A member of the Council will act as the Chairperson.

Step III: Recommendations of the Search Committee are submitted to the Selection Committee which will interview all the recommended candidates and select either one or none for the consideration of the Council. We decided that views of all stakeholders must be sought. This Committee will therefore consist of a total of 11 members. As Chairman of the Council, I became the Chairman of the Committee. The Chairman of the Search Committee became a *de facto* member. The remaining nine members were elected representatives of the various stakeholders – two members each from the teaching staff, professorial staff, non- teaching staff, student (one undergraduate and one postgraduate) and an extra elected member from the Council. All members have to abide by the rule of confidentiality and the collective decision made at the Committee meeting. I believed that we should come up with either one candidate or no candidate at all. There should not be an alternative candidate. Either the candidate is good enough for the University or he is not.

Step IV: On an appointed date, the candidate will have to meet the stakeholders one by one: academic staff, non-academics, students (undergraduate and postgraduate) and alumni. Each has to make commendations to the Council.

Step V: In accordance with the statute, the candidate will have to be assessed by the Senate.

Step VI: Council then convenes to make a decision. After hearing recommendations and comments from all stakeholders: the Senate and the Selection Committee, the Council will either approve the candidate and appoint the person and from then on give him all the necessary support he needs to carry out his vision and duties, or reject him and restart the search process.

Peter Mathieson, former Dean of medicine and dentistry of the University of Bristol was recommended by the Selection Committee. The different stakeholders showed support and the Senate approved the nomination. The Council was given the choice to either appoint him for five years or to reject him and restart the search process.

Despite being the Chairman of the Selection Committee, I did not influence the selection process in any way. I gave the following order: the Council will have to make a collective decision. If the decision is to approve the candidate, it is our power to appoint him and to give him full support to carry out his work. If the decision is to reject him, so be it, and the Council will have to restart the whole search process with a set of new committees following the same principles and procedures.

Peter Mathieson was ultimately appointed as the 15th Vice Chancellor of HKU.

Whether we had selected the right candidate under the circumstances, history will tell. I will not defend the choice. Yet I must defend our system. It is a system based on multiple and sensible criteria. It actively seeks and takes into consideration the views of all stakeholders. It is a system that is totally transparent and unbiased. It should be a system that future selection of Vice Chancellors should continue to adopt.

The search was over. The saga, however, was far from over.

As soon as the appointment was announced there was hue and cry. There were all sorts of criticism. "The candidate is not Chinese"; "He does not understand our culture"; "He does not have tip top academic credentials". Some of the comments were nothing but fantasies bordering on being ridiculous. "The Chairman recommended him because they are both kidney specialists"; "The Chairman supports him because they are acquaintances, both sit on the editorial board of a medical journal". I did not dignify these comments with any response, let alone defend my integrity. The selection process was based on established, pre-approved criteria. The Chairman remained neutral in the entire transparent process.

How ironic for these people to demand "institutional autonomy" on the one hand, and tried to influence the University's decision at every opportunity on the other! But sadly this is Hong Kong today. The Council was given the mandate to select and appoint a candidate, and it had done so by following due process and giving consideration to the views of all stakeholders. Yet some cried foul because the selected candidate was not their "cup of tea".

The search for a Pro-Vice Chancellor (Research and Manpower) was plagued by a similar problem. This time, the task was entrusted to a Search Committee headed by the Vice Chancellor assisted by three elected professors and a member of the Council. After a world-wide search, the Committee will recommend a candidate, if any, to the Council for consideration. For the protection of the interest of the candidates, the whole process must be kept strictly confidential. No names should be revealed.

Regrettably, during the search, there was a "leak" that a certain internal candidate was the preferred candidate. This was followed by intensive lobbying by certain factions of the alumni and students to have that candidate appointed. Meanwhile, there were accusations that the Liaison Office of the Central People's Government was putting pressure on the Council members. Yet any concern or worry that the best possible candidate would not be selected was uncalled for. A recommendation was finally presented to the Council. After much deliberation, the Council came to a collective decision to reject the recommended candidate on the ground that "his appointment" was not in the best interest of the University. All hell broke loose! Students stormed the Council Chamber wreaking havoc, demanding explanations and not allowing the Council members to leave. I stepped in to answer their questions and calmed them down. The "Storming of the Council" came to an end. Deliberations of the Council were supposed to be confidential. Yet this time, the deliberations were leaked to the media. The Council members, however, stood their ground and uphold their decision.

Council's deliberations must be confidential. It is only when confidentiality is maintained that the members are given the chance to express their thoughts, concerns and feelings freely and without worries or regrets. Deliberate infringement of confidentiality damages the integrity of the Council and denigrates the honour of the University. This is why every member swears to abide by this principle. This time, the infringement was so blatant despite repeated warnings that I, as the Chairman, had to act on a "chairman order" and applied for a court injunction to have the leakage banned from being reported on the media.

The atmosphere of the society today is highly politically-charged. People are divided and often times practise double standard. The Council was accused of kowtowing to the Liaison Office of Central People's Government. The accusation was unproven and unsubstantiated. Yet attempts to pressurize the Council into selecting their preferred candidate were blatant and extensively reported by the media. That decision of the Council was challenged by two students who sought leave to the Court of Law for a judicial review. Rationality, I believed, triumphed again. The leave for judicial review was not approved.

Peter Mathieson assumed office as the Vice Chancellor and President on April 1, 2014 (April Fools' Day). He resigned after serving just over three and a half years and left without completing his five-year appointment term to become the Vice Chancellor and President of the University of Edinburgh. He was appointed by me and I was disappointed, for to me making this move would leave a dent in his academic career and integrity. I was further disappointed that prior to demitting office, he made several uncalled-for remarks insinuating that the Central Liaison Office had interfered with his work. Such remarks were uncalled for, and

Having a dialogue with students disrupting a Council meeting (Courtesy of *Sing Tao Daily*)

they caused a rift between the University and the Central Liaison Office as well as the Hong Kong SAR Government. It would take time and efforts for the Council and his successor to have this rift mended.

Cooperation with the Mainland

The Faculty of Medicine has a history of over 100 years and had one of her first students Dr. Sun Yat-sen. Since inception, she has produced numerous doctors and of late, nurses and traditional Chinese medicine practitioners that serve the societies of Hong Kong, China and, to a certain extent, the world. Many have moved to engaged in public services and benefit the society in more ways than health care. Yet, surprising enough, the medical school did not even have an affiliated hospital. It was a misnomer to call the Queen Mary Hospital and the Prince of Wales Hospital the affiliated hospitals of the medical schools of HKU and the Chinese University of Hong Kong respectively. These two hospitals are granted to the two medical schools for the teaching of their medical students as well as the training of staff and research. The universities have no actual say over the control of the hospitals. When the Hospital Authority was formed, the idea was to relinquish the two hospitals to the respective medical schools. According to this idea, a determine budget would be allocated to the medical schools and each medical school would take full responsibility for each hospital. The University Grants Committee and the two universities, however, rejected the offer. They believed that the core values of the universities did not lie in managing or running hospitals and they did not have the necessary expertise. The idea was therefore dropped. Instead, a Teaching Hospital Subcommittee was formed under the Hospital Authority, consisting of its Chairman and Chief Executive, the Vice Chancellors and Presidents and Deans of the medical schools to meet regularly to discuss issues of common interest and concern.

Was this a wise decision? This arrangement relieved the universities of the burden to run a major hospital and to balance its budget. It also meant, however, the universities would continue to do without their own teaching hospitals, and had to settle with borrowed places under the control of the Hospital Authority.

In late 2000, HKU was approached by the Ministry of Health of the Central People's Government to see whether she was interested to cooperate with the Shenzhen Government to start a new hospital called Binhai Hospital (濱海醫院). This presented us with a dilemma. While we saw it as a great opportunity for HKU to "fly off" again, we were hesitant to throw ourselves into managing a hospital and especially its budget.

We finally accepted the challenge with understanding view to accomplishing the followings:

We will be assisting the motherland in her quest for health care reform by introducing a Hong Kong system into the country;

We would be expanding HKU's facilities in teaching and training when such resources in Hong Kong are dwindling and proving inadequate for an increasing student population and staff pool. There will also be more space for research; and a chance to tap into China's gigantic medical research funding pool;

We would be providing a modern tertiary hospital (三甲醫院) for the Government and citizens of Shenzhen. They will no longer need to go to Guangzhou for every major medical problem.

The "devil", as they say, is in the details. But, we took a bold step!

This Hospital, with 2,000 beds and modern facilities, would be totally funded by the Shenzhen Government (the cost amounting to 4 billion yuan and initial equipment costing the same amount of money). But if it was to succeed, it would have to be managed in the way we proposed:

- It will be a HKU-affiliated hospital (the name Binhai Hospital was changed to The University of Hong Kong-Shenzhen Hospital (香港大學深圳醫院));

- Unlike most China enterprises, the Chief Executive of the Hospital reports to a Board and not to a politically-appointed Party Secretary. In short, the management of the hospital will in no way be influenced by politics;
- Patient management is run by the senior management team of the Faculty of Medicine of HKU;
- All patients go through a primary family medicine service and only those in special need will be referred to specialist clinics;
- A fixed fee will be charged for the out-patient family primary clinic that will cover consultations, basic investigations, and basic treatments and roots out over investigations and poly-pharmacies. We totally forbid "rebate", "crawl back" and "under the table red package".

The formation of the Hospital Board again represented the triumph of rationality and the success brought by compromise. The proposal was for the board to be formed by equal number of members from HKU and from Shenzhen. The proposal also requested that it should be a "Double Chairmanship" shared by HKU and Shenzhen. I was opposed to the idea which I believed, if executed, would lead to chaos when two persons are running the show. Without hesitation, I gave up the right for the title and proposed that the Chairman of the Board should be the senior person on the Shenzhen team. After all, they are more familiar with the rules and regulations in the Mainland.

HKU was blessed with the fact that we have an extremely accommodating Vice Mayor in the person of Wu Yihua (吳以環)

With the Vice Mayor of Shenzhen, Wu Yihuan (吳以環) (the third from the right in the first row)

as the Chairman of the Board. She was most understanding and helpful, always willing to accommodate our needs, and often bending over backwards to lend a supporting hand.

The search for a Chief Executive of the Hospital was a formidable task. While there is no hard-and-fast rule to determine the quality of a Chief Executive, some of the criteria used in the Mainland included: he or she should be a medically qualified doctor; someone "known" to the Ministry of Health; and someone commanding respect in the medical circle of Hong Kong. Yes, the Chief Executive is to be nominated by HKU, but I believe as a matter of courtesy, we should take the commonly used criteria in China and the views of the Chinese Ministry of Health seriously. There were also suggestions that since the Faculty of Medicine of HKU did not have expertise in hospital management, it might be better to acquire the services of experts outside the Faculty with experience in running hospitals or hospital chains.

I was opposed to this idea. The Chief Executive will have to lead and command the respect of a team of senior clinical academics and professors of the faculty to run the hospital services. He or she, therefore, must have a thorough understanding of the likes and dislikes of his clinical team, a thorough knowledge of the members' behaviour and working habits. He also needs to command the respect of the senior members of the Faculty. In short, as the head of the team, the future Chief Executive must be a senior and well-respected member of the Faculty. There was therefore very little choice.

Prof. Grace Tang (鄧惠瓊), chair professor in obstetrics and gynecology, one-time Dean of medicine, and currently the Chairman of the Faculty Board was chosen against her wish to be nominated for the Chief Executive. It turned out that she was great at her job, despite her lack of hospital running experience.

Under the MOU (Memorandum of Understanding) of the Hospital, she led a term of four Vice Chief Executives – a Deputy Chief Executive, nominated by the Shenzhen Government whose main role is to negotiate and liaise with the Shenzhen authorities. I endorsed the concept. After all, we needed a person who understand the rules and regulations of the Mainland to get things done in Shenzhen and the Mainland. The other three Vice Chief Executives are responsible for looking after clinical services, supporting services, and, as this is a teaching hospital, research. They were nominated by HKU.

The key role of any hospital is to provide clinical services, care and treatment to the sick and the destitute. How this goal is

Celebrating the opening of the HKU-Shenzhen Hospital (Courtesy of the University of Hong Kong)

The HKU-Shenzhen Hospital

accomplished varies from hospital to hospital and from region to region. In the case of HKU-Shenzhen Hospital, the basic aim is to introduce the Hong Kong management and the clinical services concept into the Mainland. In essence, each clinical unit of the Hospital should be led by three senior HKU heads managing a team of local staff.

Three professors and senior members of the clinical team, therefore, must work in Shenzhen on a daily basis – would that have an adverse effect on the service the University was providing in Hong Kong? And why should money from Hong Kong public coffers be spent on Shenzhen?

Three extra full-time equivalents (FTE) would therefore have to be provided for each Department so as not to undermine the service standards of Hong Kong. The financial burden will have to be borne by the Hospital, initially paid upfront via HKU.

Tides of criticisms, objections, oppositions and negative comments swept from all sides, even within the Council of the University itself. There were serious concerns over whether the Hospital could balance its budget.

But it was also a great opportunity for us to live up to our motto that HKU is for Hong Kong, China and the world. What is equally true is that HKU stands to gain many obvious and hidden benefits from running a hospital in Shenzhen. Again, after weighing all the pros and cons, the Council made the right decision.

The Hospital just celebrated its 5th anniversary. Was it a triumph or a failure? Well, we are now taking care of over 7,000 outpatients a day. We get the Australian Council on Healthcare Standards (ACHS) accreditation within the shortest possible time and we acquired China's tertiary hospital status. What's more, we have made some significant changes to the local culture: getting rid of the unnecessary "IV drip room"; providing our staff with a safe working environment by forbidding patients to abuse staff. A public security official is stationed in our Accident and Emergency Department. To put the minds of staff at ease in the pursuit of their duties, they are covered with adequate practice indemnity. The practice of accepting red packets was totally abolished. In fact, staff now abhor the practice as they are taking home a reasonable pay package. It is now their belief that serving their patients is their duty and not a means of getting an extra income.

Were there any shortfall? It would be foolhardy to expect running a 2,000-bed hospital with a view to introducing hospital management reforms to be plain sailing. Yes, there were setbacks and failures: we have not been able to put in place a centre for organ transplantation; we have not been able to deal with open heart surgery and we have not been able to accept all emergences brought in by ambulance service. It is because we still have not complied with the set regulations of each situation. Our private patient attendance is still below par, and thus is affecting the financial health of the Hospital.

The media, in particular, the Hong Kong media is always ready to point an accusing finger at us. They leave no stone unturned in their attempt to tarnish us or to expose our deficiency, especially concerning our budget and service standards.

I am glad to say that our balance sheet is totally transparent. As for deficit, it usually takes years for a corporation of our size to balance its budget.

An independent survey and assessment carried out by a reputable independent firm (Price Waterhouse Cooper) indicates that barring any mishaps, the Hospital will balance its books by 2018 and fully repay HKU's upfront funding for FTE several years later.

My salute to the management, and the staff of HKU-Shenzhen Hospital, for how they often go beyond their call of duty to serve the patients and do their jobs well.

I was once asked what the most difficult problem the hospital had encountered so far was. Since local staff members came from so many different parts of China, did the mix of clinical experiences and expertise give rise to any problems? My answer was simple. The professional standard of the staff from the Mainland was high, occasionally even higher than those of Hong Kong staff. It is the cultural differences that need to be managed. These include the cultural differences of the health care staff, the Shenzhen Government, the Board as well as the patients. The Hong Kong system is based on commitment and transparency. The medical professionals take pride in their jobs. There is no shortcut and everything must be done by the book.

Local staff members of the Hospital came from cities and provinces all over China. It is therefore only natural that there are different approaches to treating patients with even the same medical problems. But to ensure an effective and efficient service,

Having a taste of the Hospital's general ward service (Courtesy of the University of Hong Kong)

the Hong Kong system must be adhered to. Patients, for example, are not allowed to choose their doctors and definitely not by giving hospital staff "red packets".

Patients must be assessed by a primary-care physician first before being referred to a specialist. This practice does not only save precious resources, but it also spares patients unnecessary investigations. The common practice of setting up an intravenous line unthinkingly is not only discouraged but also condemned in our hospital. They work in an environment and culture that gives priority to the interests of the patients.

In line with the policy to discourage corruption by high pay, staff members are reasonably well-paid, especially compared to their counterparts working in other hospitals in China. When the salaries of medical practitioners are not commensurate with their expertise, they often succumb to the temptation to accept bribes or rebates. Medical practices, therefore, become financially-driven, leading to a decline in the professionalism and social status of the medical profession. Gone are the days when even the chauffeur of high-ranking government officials can exert influence and claim priorities to have their relatives and families admitted for treatment. Admission of patients and who will be treated first are based on the severity of their conditions and therefore purely medical grounds, not because of the positions of power of their relatives.

Changing cultures is never easy. Old habits die hard and people are adverse to change. In the case of accepting "red pockets", for example, patients themselves often get upset if their "gifts" are refused. They think they are not being taken seriously, and therefore, will not be given the medical treatment they need.

Successful treatment depends very much on the "doctor-patient" relationship which is built on trust, not financial incentives or benefits.

During the seven years when I took the helm of the HKU-Shenzhen Hospital, I did my best to introduce to the motherland not only a better, more sophisticated health care system, but also one that better serves the interests of both the patients and the medical professionals. For me, this fulfilled a dream I had harboured ever since my first encounter with China's medical system. China's medical professionals are renowned for their innovation and the richness of their clinical experiences cannot be matched. Yet unless they have the chance to work under a more rational and well-designed system, they will find it difficult to go to the next level and compete successfully with the outside world.

My years with HKU-Shenzhen Hospital were in many ways an educational experience. I was most impressed by the commitment and dedication of our health care professionals, the understanding and accommodation of the Shenzhen Government, and the trust the two sides have in each other. The hospital staff from Hong Kong travelled to and from Shenzhen and put up with great pressure in an unfamiliar working environment. The Shenzhen Government, in particular the Vice Mayor, had absolute confidence in our concept and our system. She often bended over backward to give us what we needed.

To her, I offer my sincere gratitude and my utmost respect. I met with her many times in Shenzhen to share our ideals for the hospital and health care services, always in the presence of Isabella Wong (王依倩) as my Putonghua was and still abysmal.

As my term of office as Chairman of HKU Council (after servicing the Council for over 11 years – more than five years as Council Member and another six years as Chairman) came to a close, I also demitted office as Honourary Chief Executive of HKU-Shenzhen Hospital and member of the Hospital Board. As a matter of principle, I made it a point not to comment on the policy of any institution which I had served after my departure. Nobody is more annoying than a backseat driver. Besides, your views and thoughts may not go down well with the new management in a new environment. More importantly no one is indispensable. I left the Hospital with mixed feelings. The Hospital had come a long way since its opening and I am confident of its success in the future. It saddened me, however, to leave an institution I helped to found and have to say goodbye to the officials of the Shenzhen Government who had become my friends. I will never give up that friendship. I left with tears in my eyes. The Board very kindly presented me with a work of calligraphy as a gift.

Expansion of HKU

The "campus" of HKU at Pokfulam near the Western District of Hong Kong is located in a heavily populated area where property prices are extremely high.

As the University continues to grow, much more space is needed to house the staff and students, and for the development of research and teaching facilities that a growing university requires.

In the west of the main campus, up on the top of a hill, and across the Belcher's was a piece of land occupied by a reservoir's filter bed. This piece of estate would be ideal for the building of a new campus facility to coincide with the celebration of the University's Centenary.

The idea was thus conceived. The plan was to move the filter beds into a tunnel in the adjacent hill, leave behind a piece of flat land for the Centennial Campus.

It would be quite an engineering feat. The whole filter bed was ultimately moved into the excavated base of the adjacent hill. This may remind the more mature readers, of the Hollywood movie *Guns of Navarone* in which Germany tanks are hidden in a mountain. In 2012, three faculty towers – Faculty of Law, Faculty of Arts and Faculty of Social Sciences – were moved to the site.

Such rapid expansion wouldn't have been possible if not for the generosity of donors whose passion for education and faith in HKU I find so genuine and touching.

The late Lady Mona Fong (方逸華), wife of the late Sir Run Run Shaw (邵逸夫) of the Shaw Foundation was grateful to Prof. K.Y. Yuen (袁國勇), a microbiologist in our own Faculty of Medicine and an infection disease expert for looking after her late husband when he was ill. The Shaw Foundation had already made many pledges for donation to HKU.

On that basis, I approached her with a plan of the Centennial Campus and offered to name one Faculty tower after Sir Run Run Shaw. Without hesitation, she agreed to make a donation. It was indeed an exceedingly generous donation but still fell short of our expectations. I boldly asked for a more substantial amount. She responded by saying she would make a decision by the evening. Around 8pm I received a call from her. She agreed. It had nothing to do with my persuasiveness, but everything to do with her faith in education and love of the University.

It is a fairly typical example. Throughout these 100 years, HKU has been blessed with having many philanthropists, starting with Sir H.N. Mody, a major donor towards the founding of HKU. He was followed by the likes of Mr. Li Ka-shing, Lord Kadoorie, Mr. Cheng Yu-tung (鄭裕彤), Mr. Li Shau-kee (李兆基), Mr. Stanley Ho (何鴻燊), The Jockey Club Charities and many others. To all of them the University is most grateful.

Every donation has a story. Mr. Tam Wah-ching (譚華正), for example, sold some of his properties to create a personal chair in dentistry and to donate to modernize the University's main library.

Ms. Nancy Chau Sing-wo (周善和), a single lady from the Mainland, was hardly educated. She worked variously as a manual labour, house maid and elderly home helper. She saved almost every "penny" she earned and donated a large part of her savings to the Faculty of Medicine to assist medical students with financial difficulties to complete their studies.

HKU students are the beneficiaries of such selflessness and generosity, without which they can never rise to such prominence after graduation. That's why they should be grateful and make use of every opportunity to repay their debt by serving the people the best they can.

The opening of the HKU MTR station on December 28, 2014 was another milestone of the University. The station was part of the MTR's West Island Line, a westward extension to its existing Island Line. The largest and deepest station in the MTR network, the station located near HKU, is an engineering marvel. Starting from below sea level, it moves upwards to mid-level in a few seconds and opens into two portals – the University's old wing

A touch of nostalgia – My father (Standing ten from left in the old photo behind us), my brother and me at the HKU Station

and the Centenary Campus. It also testifies to the unique place the University occupies in Hong Kong and the enormous respect her people have for the institution. Yes, before the opening of the HKU MTR station, there were MTR stations in the vicinity of other universities in Hong Kong, such as the University Station near the Chinese University of Hong Kong and the Kowloon Tong Station near the City University of Hong Kong. But no other MTR station in Hong Kong is named after a university. That station, in the vicinity of HKU, was named "HKU Station" (香港大學站). Needless to say, the then MTR Chairman Mr. Jack So (蘇澤光) is a HKU alumnus.

When you get off at the HKU station, walk along the hallway and take the assembling elevator, you will not be showered with advertisement of “skin care” and “body beauty”, instead you will get a taste of history – pictures of HKU over the past 100 years, from Dr. Sun Yat-sen to the modern day, from the laying of the foundation stone by Sir Fredrick Lugard to the opening of the Centenary Campus.

If you watch these pictures closely enough, you might spot my father, myself and my brother amongst them. After all, HKU is our *alma mater.*

Chapter 10

It is all about compromises

It is often said that politics is the art of the possible. This is probably true. To be a successful politician, you have to turn the table by making the impossible possible.

No two persons can see exactly eye to eye about the same thing. If a deal is to be made, both sides have to behave rationally, understand the importance of the outcome, give in a little to the demands of the other side and let rationality triumph. This is what we call compromise. Any disagreement on details and technicalities can always be dealt with later. The Chinese dictum "seeking common ground while retaining small differences" (求大同存小異) remains the best way to achieve success in negotiation.

Labour dispute or confrontation between employers and employees is a universal problem. It is also a perpetual problem. In many cases the disputes are such that there could be no end in sight.

Employees want less work and more pay, while employers want the opposite for workers.

In Hong Kong, the bones of contention are: minimal wage, the hedging mechanism of Mandatory Provident Fund Scheme, statutory holidays, importation of foreign labour, collective bargaining for employees, and standard working hours.

Employees believe that they are exploited by their employers and are unable to share the fruits of economic success. They think they are entitled to not just better pay but more statutory holidays. They are against the importation of foreign labour for fear that they will be priced out of the market. Today, when everyone is talking about work-life balance, many of them are struggling with long working hours that take a toll on their physical and mental health. If what they do concerns public safety, such as bus driving, they become a potential danger to the society. It is a common concern too, that long working hours rob employees of their family lives. But the truth is standard working hours alone does not guarantee employees shorter working hours, unless there is also a control on maximum working hours.

Ironically, it is the labour unions, not the employees themselves, that clamour loudest for standard working hours. This is not surprising. The largest trade unions in Hong Kong are affiliated with political parties, such as the Federation of Trade Unions (香港工會聯合會), Hong Kong Federation of Trade Unions (香港職工會聯盟) and Neighbourhood and Worker's Service Centre (街坊工友服務處), etc. They are all jockeying for members and are therefore keen to be seen fighting for the welfare and benefits of workers

Employers usually exercise greater restrain. Yet they are firmly and vehemently opposed to any form of working hour regulation. As they see it, such legislation will incur extra cost especially for small and medium enterprises (SMEs) and accentuate their manpower shortage, especially when the importation of foreign workers was not allowed at that time. And since SMEs account for over 80% of Hong Kong's companies, imposing standard working hours, they believe, will damage Hong Kong's competitiveness and the economy. They also worry that any legislation, if passed, will be subject to regular review with the working hours further shortened with each review.

The issue is thus a political hot potato. When I was asked to chair a committee on standard working hours, I was therefore hesitant. I was not familiar with labour disputes. I run my own clinic with only two nurses and one secretary. These weaknesses turned out to be my strengths as they exempted me from any conflict of interest. Perhaps that's why both the employers and employees found me "acceptable".

The terms of reference of this Committee were as follows:

- To follow up on the Government's policy study on standard working hours and conduct further in-depth studies, as necessary, on the key issues identified therein;
- To promote understanding of standard working hours and related issues including, among others, employees' overtime work conditions and arrangements; to engage the public in informed discussion on the relevant issues, and to gauge the views of stakeholders; and
- To report to the Chief Executive and advise on the

Mobbed by a crowd of labour unionists after a meeting

working hours situation in Hong Kong, including whether a statutory standard working hours regime or any other alternatives should be considered.

While agreeing to take up the challenge, I insisted on the following two directions:

- The membership of the committee must consist of the same number of employees, employers and representatives of labour unions and employer's federation. There should also be a fair number of "neutral" persons and academics to mediate discussion;
- The Government would not have a pre-conceived stance on the issue.

I knew I had an onerous, if not impossible, task. My only hope was both sides would be reasonable and let rationality triumph at the end.

At the very first meeting, members from both the employee and employer camps agreed on the following objectives and principles:

- Conduct broad-based public consultation to collect wide ranging views from all stakeholders, including employees and employer unions, employee and employer federations, major professions and the public. After all, "working hours" affects every person – shorter working hours may mean less income and longer working hours may mean less time spent with one's family;
- Agree "to seek common ground while retaining small differences";
- Whatever conclusions we reach, they must be based on evidence;
- Any decision we make should not affect Hong Kong's competitiveness and economic development; and
- Due regard must be given to both the protection of employees' rights and the affordability of enterprises.

We conducted many rounds of independent, broad-based consultations consisting of the followings:

- Over 10,000 successful face-to-face household surveys, interviewing employees at random about their actual working hours based on diary entries;

- Small-group meetings with labour unions and employer federations;
- Survey of select professions and occupations;
- Extensive public consultation through open forums.

When the results were analyzed and presented to the Committee, I thought I hit the jackpot. There was consensus amongst members on at least four areas:

- There should be legislation on working hours policy;
- There should be a legally binding contract between employers and each employee governing working hours, overtime pay, rest periods, etc.;
- Since each job, each trade and each profession works differently, a "one-size-fits-all" approach to regulating working hours is not feasible;
- Greater protection should be provided to grassroots workers with poor negotiation power.

The optimism, however, proved premature. No sooner than the Committee launched its second-phase consultation than the labour representatives on the Committee decided to walk out. They boycott the Committee and the second-phase consultation, declaring that they would only return if the Committee agreed to legislate on standard working hours. Never mind the initial agreement we reached.

We were being held to hostage. As the Chairman, I had two choices: to dissolve the Committee and hand the work back to the Government; or to stand our ground, finish our job and submit our recommendations.

With the support of the remaining members, we moved on. After all,

- we had enough members to form a quorum. We should not scuttle the Committee because a few members were absent;
- we had members on the Committee who represented employees though none could speak for labour unions;
- we had, through the first round of consultation, collected a fairly sizable amount of views from employees; and
- the labour union representatives essentially gave up their rights to express the views of their members and betrayed their obligation to fight for their interests. Results of the negotiation can only be achieved at the meeting table.

The Committee persevered, and completed the second-phase consultation. In this exercise we tried to identify a consensus or at least a dominant view as to what level of employment to which "standard working hours" should apply.

We used three basic parameters for analysis: salary level, working hours and overtime compensation.

It is a common sense that senior and management-level employees cannot be subjected to working hours regulations. They may be "off duty" after working hours, but their commitment to their work and responsibilities are open-ended.

Obviously standard working hours can only apply to those employees with lesser responsibilities. But how do we identify these employees?

At a consultation session

Smiles can be deceptive – with Stanley Ng (right, labours' representative) and the late Stanley Lau (left, employers' representative)

Regrettably, despite some 40 sessions of intensive consultation with various stakeholders, not a single recommended proposal emerged. While employers refused to suggest a salary level for fear of being ridiculed by labour unions, employees insisted that standard working hours be extended to all jobs and all salary levels. The result was a total impasse. This intransigence could be attributed to the cut-throat competition between the many trade unions in Hong Kong for members. They were quick to point an accusing finger at one another. Any compromise on the issue of standard working hours, no matter how reasonable, could be called "selling the workers down the river".

Meanwhile, the labour unions conducted their own surveys and demanded the followings:

- Legislate standard working hours for all;
- 44 hours per week as standard working hours for all;
- Overtime be compensated either by time off or at the rate of 1.5 times regular hourly salary.

In their consultation report, they also came up with two slogans:

- "Work with no end" (有開工冇收工);
- "Overtime with no compensation" (有超時冇補水).

I grasped the opportunity, and came up with two proposals that addressed their complaints:

- All employees and employers must sign a legal contract in relation to working hours, etc. This will put an end to work with no end;
- Any overtime work must be compensated by statute at the

Speaking at a consultation session (Courtesy of Mr. Ducky Tse)

rate of their existing hourly pay or more. This will tackle the issue of overtime with no compensation.

As expected, the labour unions were not impressed. They refused to accept it for what we proposed was not a piece of legislation on standard working hours, but a law for contract working hours. The irony is that as of today, they have never objected to these two proposals and there has never been any hue and cry.

In June 2017, the Government announced that the Chief Executive in Council had endorsed the Committee's report and recommendations as "a general framework for guiding future formulation of the working hours policy" and adopted "suitable supplementary measures to take forward its recommendations".

For more than three year and after having conducted a total

of 80 consultation sessions, and detailed analysis of the needs of both employees and employers and the economic impact of working hours policy on Hong Kong, we were not able to reach a compromise. Yet an important first step has been taken, and the essential groundwork for regulating working hours has been laid.

It was also an educational experience for me. I realize now there is a small number of "unconscionable employers" in Hong Kong whose exploitation of their workers can be shameless, though many are honest, fair and caring. It also gave me insight into the working culture of Hong Kong employees which is different from that in the West. As the veteran labour activist Lau Chin-shek (劉千石) told me, "employees from the West work for holiday. Hong Kong employees work for salary".

The Committee's work was also valuable in two other aspects. We have collected a sizable amount of data on the actual working hours attitude and behaviour of over 10,000 workers from different trades. With the help of the Census and Statistics Department and the Economic Analysis and Business Facilitation Unit, we have conducted an in-depth economic impact analysis based on salary levels, fixed working hours and graded overtime compensation.

Hong Kong will face an increasingly mounting shortage of manpower after 2018 when her labour work force peaks. At the same time, Hong Kong is becoming a more and more service-oriented economy. How will standard working hours affect the big picture of Hong Kong's competitiveness under such circumstances? The data and analysis the Committee provides will help us answer the question.

But that is the picture. The issue of standard working hours is emotive that arouses people's passion. As the Chairman of the Committee, I often found myself the subject of attack and ridicule. There were insults, character assassination and people openly called for my resignation. Once I was mobbed by a crowd blocking me from leaving the meeting venue unless I consent to legalize working hours. But I wasn't intimidated and I held on to my principle. In the end, it was the security guards and the police who got me out. To them I offer my personal thanks.

At another time, I was requested to attend a meeting of a labour union. Against the advice of the staff of the Standard Working Hours Committee, I decided to go. I saw this as an opportunity to drive home my message of rationality and conciliation and to extend an olive branch to my detractors.

Despite occasional slogan shouting and placard demonstrations, the meeting otherwise went smoothly. When the whole questions and answers session was over and the convener announced the end of the meeting, a senior official of the union came up and offer me a "present" – a diaper which I accepted with grace. I was then asked to unfold it and I did. I was in no way disturbed and I said with a smile, "I am an urologist. Human waste is something I encounter every day!"

I relate this incident here not because I hold grudges. It gives readers an idea of what Hong Kong has become. Such intolerance and indecency have come to define the type of democracy some people practise.

Epilogue

The triumph of rationality

This book is not an autobiography.

The stories and events described here reflect the lives and times of the people of Hong Kong during those turbulent years leading up to and shortly after the change of sovereignty.

I myself was no more than a "pawn" swept by the waves of change. Yet, at the same time, I was also a player in a melodrama who, like a "white knight", "slew the villain" and rescued the "fair damsel".

I was blessed with the gift of vision – able to see what was coming – and I did my best to seize the moment and make a difference. This had enabled me to effect positive changes in a range of issues. This also led to the various efforts, detailed in the preceding chapters, that I had made to make the medical profession more autonomous and to help Hong Kong become a more caring, friendly society, especially for the elderly and the workers.

As the song, *My Way*, goes, "Regrets, I've had a few. But then again, too few to mention". Like everything else, success never comes easy. Often times, stakeholder are resistant to changes. Yet, rationality always triumphs, in one way or another and under different circumstances. It always stands the test of time. The eventual recognition of traditional Chinese medicine practitioner (TCMP) through statutory registration is an obvious example. I was pushing for a law to register traditional Chinese medicine practitioner, yet I faced stiff opposition from all sides – not least the practitioners themselves who saw a doctor trained in Western medicine interfering with their practice and seemingly attempting to damage their "rice bowl". I also faced opposition from my own constituents. This was not hard to understand. If traditional Chinese medicine practitioners were allowed to practise as health care providers, it would increase medical manpower and cut down drastically the population to medical manpower ratio.

This is to put personal interests before the interests of the society. Yet proper registration was a way to regulate traditional Chinese medicine practitioners and would provide greater protection to the public. It was also the first step towards developing traditional Chinese medicine which has played a role for the health of Chinese people, be they royalties or commoners, for thousands of years but has been left in neglect for some 100 years. At a time when the whole world is advocating "herbal" medicine, should we, as descendants of the Yellow Emperor and Hua Tuo (華陀), do nothing and let traditional Chinese medicine vanish into oblivion?

I might be over zealous when I tried to include traditional Chinese medicine practitioners in the "medical functional constituency" of the Legislative Council. After all, as health care providers, we should speak with one voice for the improvement of health care services and to fight for more resources from the Government and the public medical service. Regrettably, personal agenda, as always, could masquerade public good. My view was not shared by many of my Western medicine-trained colleagues who were up in arms, fearing that one day a traditional Chinese medicine practitioner would represent them in the Legislative Council representative role. What short-sightedness and lack of confidence! As expected, my idea was scuttled.

Yet, again, rationality triumphed, traditional Chinese medicine practitioners are now under the well-functioning registration system and traditional Chinese medicine is developing in Hong Kong. There are now 18 public traditional Chinese medicine clinics providing health care services to the public; three of our UGC-funded universities have opened full-time Chinese medicine practitioner courses conferring bachelor degrees and a stand-alone traditional Chinese medicine hospital is in the pipeline. Traditional Chinese medicine practitioners may one day run against the Western medicine-trained practitioners in election, instead of working together for their common interests. It was an opportunity missed.

As technology advances and the society changes, old values may be discarded and long-standing conventions ignored. Assisted scientific human reproduction procedures are a case in point. Yes, science can improve our quality of life, yet unchecked scientific progress can wreak havoc on society. Science is a miracle worker to those who cannot conceive. It can bring to the world a physically normal baby to everybody's satisfaction. But the ethical, social, moral and legal issues involved must be carefully considered to protect the interest of the child conceived under such circumstances, as I take pains to point out in the chapter on doctors and society.

Artificial insemination, or achieving pregnancy from "borrowed' sperms, seems like a straightforward scientific procedure. A child is born yet this offspring is forever a "bastard". Laws must therefore be enacted to stipulate that if the husband has given consent to the procedure of artificial insemination for his wife, he will be, both morally and legally, the father of that child. The laws of Hong Kong and Chinese tradition define a married couple as a man and a woman or husband and wife. In many parts of the world, same-sex marriages are now legal. But just imagine the hassle a child has to go through when asked who his or her parents were. I have nothing against same-sex marriage, but the children of same-sex parents may be forever stigmatized in a conservative society. Then consider a situation in which a mother offers to be the surrogate for her daughter who is unable to carry a baby. When the child is born, who will be his or her mother, sister or grandmother?

In Hong Kong, the legal age of consent to donate a life organ is 18. There is nothing arbitrary about this number. A person below 18 is considered not mature enough physically or mentally to understand the implication of donating an entire or a part of his organ and its possible long-term harmful effects. He or she could be coerced. The request may take the form of emotional blackmail – "if you do not donate, your mother/father will die". It is for the same reason that a child below 18 is not given the right to vote at elections. Put yourself in the shoes of the donor's parent whose underage, perfectly healthy offspring, has to undergo a major surgery with uncertain outcome and possible long-term side effects. And put yourself in the shoes of the surgeon, who has to use his scalpel to remove an organ or a part of an organ from the body of a healthy underage child. What psychological stress will he be under? How devastated will he be if he makes even a minor mistake and ends up losing not just the donated organ but the donor as well? After all, the reported overall complication rate of living organ donation is about 20%.

It is very easy for those who do not have to make these agonizing decisions and go through these psychological tortures to support an amendment lowering the age of consent for organ donations. After all, who doesn't want to be seen as a life-saver? The amendment was in fact seriously considered by the powers-that-be and, if introduced in the Legislative Council, was very likely to have the backing of the lawmakers. Fortunately, rationality again triumphed, and alas the amendment was not introduced!

Following the example of my father, I decided to pursue a career in medicine at the age of 23. Surgery was my chosen profession, as the process of performing surgery involves making quick decisions and achieving positive results. It turned out to be a wise choice – I obtained specialist status in three years. I was all set to become a brilliant surgeon and perhaps even a master craftsman.

But there were other things on my mind. I was unhappy with the many problems besetting the city's health care system. Instead of burying my head in the sand, I decided to do something about it. First, I became the President of the Hong Kong Surgical Society. Then I took the helm of the Hong Kong Medical Association, the main medical body that included all registered doctors of Hong Kong. I also ran for the Presidency of the British Medical Association (Hong Kong Branch) and won. I used these positions as a platform to push for my ideals to improve Hong Kong's health care policies and standards. I campaigned for, and won a seat in the Basic Law Consultation Committee. My plan was to join hands with other professionals to fight for profession autonomy after the change of sovereignty. I won a seat in the Legislative Council on my second attempt and continued to represent my constituency for a total of 12 years. For all these years, I had used my own expertise and the collective wisdom of my constituents to try to make our society a better place.

I have never defined doctors' interest narrowly. As their leader and representative, I tried to serve as the bridge between their interests and the needs of the society. Whatever is good for the society must also be good for doctors! Every time I stood up for doctors, I did so also for the society's good. I had pushed for laws that prevented industrial accidents and provided workers with proper compensations should accidents occur. Laws on pneumoconiosis and noise-induced deafness were obvious examples. I fought for laws on banning smoking. Working with Dr. Judith Mackay, and the late Prof. Anthony Hedley, I drafted and introduced to the Legislative Council the first private members' bill to curb smoking. Today, much work has been done to protect not only smokers but the victims of second-hand smoking. I believe that ultimately smoking should be banned!

One of the most controversial actions that I took was to introduce a universal licensing examination for all non-local medical graduates before they could be registered to practise medicine in Hong Kong. Public interest is at stake here. For who would like to entrust their life to the hands of a doctor unless we are assured that he at least complies with a basic standard? With thousands of medical schools around the world, how are we to judge their standards and the standards of their graduates? The most effective way must be to have these graduates sit for a basic examination. This is not, and could never be, a measure taken to protect the interest of the local graduates. In the old colonial days, British and the Commonwealth medical schools were assessed by the General Medical Council of the United Kingdom, and the Hong Kong Medical Council, of course, deferred. Graduates of other universities, including the Mainland universities, had to sit for an examination. With the return of Hong Kong to Chinese sovereignty, such an arrangement was no longer acceptable. What

Hong Kong needed was a universal licensing examination for all graduates of medical schools, which is what we have now. This examination must therefore be the only effective standard ruler. Students who were studying in the approved universities before the change of sovereignty, though, will not need to take the universal licensing examination.

Again, rationality triumphed and with the priority given to public interest.

When China's paramount leader Deng Xiaoping (鄧小平) declared that Hong Kong would be returned to China in 1997, and that was non-negotiable, all hell broke loose. Hong Kong people were making all sorts of efforts to emigrate and to close their businesses, even for those without the necessary financial means. They were hit even harder when realized that they would have no say in how the transfer was to take place as well as the terms and details of the arrangement. In other words, they were totally at the mercy of Britain and China. The introduction of the concept of "One Country, Two systems" did provide a degree of reassurance. The capitalistic system of Hong Kong could remain in the midst of a communistic or a socialistic system of the Mainland. The devil of course was in the details and how the Basic Law would be drafted.

As Hong Kong is a society that is governed by the rule of law, its laws had to be changed to tally with the "Basic Law" which would become effective immediately after the handover. This became the tall order of the Provisional Legislative Council.

I was a member of the Legislative Council from 1988 to 2000. From 1995 to 2000, I chaired the Legislative Council's House Committee. My role was to negotiate with the Chief Secretary on the Legislative Council affairs. My term of service spanned from 1988 to June 1997 under the rule of the British colonial Government, and from July 1997 to June 2000 when Hong Kong became an SAR of the Chinese Government. From July 1996 to July 1997, as a member of the Provisional Legislative Council, I participated in debates on bills and passed them into laws every Wednesday. Every Saturday morning, I travelled to Shenzhen as a member of the Chinese provisional legislature to debate and vet laws for the future Hong Kong to tally with the Basic Law.

The Provisional Legislative Council had an onerous task at hand. It had to play a dual role: to monitor the Governments' policy, and to enact laws for the current functioning of Hong Kong and to prepare for the future. At the same time, we had to amend the existing laws to tally with the Basic Law after the transfer of sovereignty. The problem was, as members of the provisional legislature, we have no mandate from the public for we were appointed by the Central People's Government. Furthermore, we were asked to confine our attention to the most necessary and urgent issues. But with or without a public mandate, we had the responsibility to ensure that the Government functions in a way that serves the best interest of the public. We simply did not have the luxury to pick and choose issues. Caught in such a "Catch 22" situation, the Provisional Legislative Council could be called "neither fish nor fowl" (非驢非馬).

The transfer of sovereignty on the midnight of July 1, 1997 was as smooth as it could be. The British retreated with grace while the Central People's Government took over with dignity. There was no commotion; in fact, most people in Hong Kong welcomed the "return to the motherland".

But that was the easy part. Are Hong Kong people ready and competent enough to govern themselves? That was the real question. The following was an excerpt from my valedictory speech in the legislative council as Chairman of the House Committee that I gave on June 27, 1997, a few days before the end of the British rule of Hong Kong:

> *Mr. President, it has frequently been said that Hong Kong people are politically inert. This may well be so, but how much is this the result of the colonial style of education that the Government has imparted to the people of Hong Kong; an education which produces perhaps good, law abiding citizens, no doubt but also an education system that dampens the mind for independent thinking; an education system that imparts the gospel of "the Government knows best, do not question it."*

Regrettably, Hong Kong people have never been given the taste of administrative participation. For more than 100 years, the concept of democracy, the concept of having people with the public's mandate to decide on policies for themselves, has never been put into the Hong Kong soil. And if it is not for the imminent transfer of sovereignty, Hong Kong people may still be in the era of "father knows best". Yet, regrettably, even with the chance some 10 years ago for granting Hong Kong people a faster step towards representative government, this was truncated. Should history be re-written otherwise, we would now have perhaps a smoother road towards "Hong Kong people ruling Hong Kong."

Since I graduated from medical school, I have been working as a doctor in general and a surgeon in particular. I have never given up this profession and I intend to keep doing it as long as possible. At the same time, from the early 1980's, I have also been heavily involved in public service, while continuing my medical practice. How could I serve two masters? I believe it is a matter of commitment, good time management, and knowing how to set priorities. Surgery helps my public services because it reins me to make rational decisions, to take calculated risks and to understand the pain and suffering that is part of the everyday life of ordinary people.

A doctor encounters patients from all walks of life. They entrust you with their lives and are therefore honest with you about their problems. By knowing these people, you arrive at a more intimate understanding of the many problems of the society and how people try to cope with them. You become more down-to-earth as you mingle not only with the rich and the famous, but also the poor and the destitute who complain to you not only their physical sufferings, but also the sufferings the society inflict on them. In this sense, taking up public services makes me a better doctor. There is great wisdom in this old Chinese saying that has inspired medical practitioners and politicians alike:

- A great doctor treats the nation (上醫醫國);
- A good doctor treats the patient (中醫醫人);
- A mediocre doctor treats the disease (下醫醫病).

From a medical craftsman, I metamorphosed myself into a technocrat, a politician and a devotee to public service. I have frequently been asked, "how many blessing could I count" and "what were my regrets". I have never believed in blowing my own trumpet. Honours and recognitions are for others to give, and history will have the last word. A life in public service and in the spotlight entails sacrifice. The media regards you as fair game and put you under the microscope. But I feel blessed that I have been able to learn continuously from the people I work with, and I take solace from the fact that my leadership of some of the most vital institutions in Hong Kong has helped to make the city a better place. I am also blessed with the understanding and support of my family, and in particular, my wife, the Founding President of Hong Kong College of Radiologists who is always there to keep me on my toes and make me a better man.

The key to successful public service is never to have a personal agenda. One should never try to have his name "engrained in a shining brass plate", because it is never about you. Furthermore, name and fame "blows away with the wind". But if you are lucky enough to have worked with the best people and if you have worked with total commitment, through the triumph of rationality, the legacy of your contributions will benefit people for years to come, despite the fact that you or your name may be left in oblivion.

My partner, my confidante and my wife Lilian
(Courtesy of Mr. Ducky Tse)

Appendix 1

McFadzean Oration

Delivered on October 23, 2004

by Dr. C. H. Leong, President, Hong Kong Academy of Medicine

Mr. President, Ladies & Gentlemen,

It is indeed a singular honour to be invited to deliver the McFadzean Oration. It is an honour to me in particular as my early medical related days was spent in McFadzean era, both as a student of medicine and as his house staff. In many aspect he was thus my mentor. No, Alexander James Smith McFadzean did not teach me surgery, in fact surgeons were a bred of doctors he dislike, but he taught me much more, as this oration would unveil.

My very initial impression of Professor McFadzean was a disarray, to say the least. As a student I was asked by him to close the door of the lecture hall, "from the outside", I was told. As a young doctor burning with desire to be a surgeon, I was baffled by why during every single Christmas, there were bound to be drawings on the wall of Queen Mary Hospital of the Professor of Medicine crossing swords with the then Professor of surgery.

Yet as I came to understand this fiery Professor from Scotland more, I start to appreciates the way he dwells into details of patients' illnesses, the environment surrounding the sickness, the societal impact the illness will bring about before exhibiting his evidence based treatment. I started to respect the way that he would stand by the profession, using all his ability to stand up to the integrity of the profession and the institute he serve. Yet he would never give in to any nonsense.

It is on these two aspirations of Alec McFadzean that I am using tonight's oration to pay tribute to.

Professor McFadzean came to benefit Hong Kong as a young man at the age of 34. Prior to this he had serve in the Middle East and Africa in an era of infectious disease – malaria, plague, tuberculosis and small pox. Between then and now many of these deadly infections have either been proclaimed to be eradicated or at least very much controlled. Small pox for example was declared eradicated in December 1979. Plague was at least temporary declared controlled when the last officially certified human case appeared in Kanataka state in India in 1966. T.B. was considered a treatable condition since Koch discovered the cause, and Malaria should NOT be fatal any more since quinine was discovered. So convinced that infectious disease were on the way out that a Harvard Public Health Group headed by Christopher Murray forecasted in late 1960 that more than 85% of all death in the US by the close of the 20th century would be due to chronic diseases such as cancer and heart problems. Rightly so, in 1900 nearly 800 Americans out of 100,000 every year died of infectious disease and by 1980 only 36.

Take another issue, of some 1,240 new drugs licensed in the 20 years after 1975, only 13(1%) were for infectious diseases primarily affecting the tropics and poor countries.

It comes as no surprise that for some five decades priorities for infectious diseases, thus public health, was dwindling. In Hong Kong before SARS struck there were only 60 infectious disease beds for a population of some seven million. In the record of the Hong Kong Medical Council's specialist registry there are only six registered as infection disease specialists. Even amongst our community medicine experts, the majority are specialists in administrative medicine, not in public health.

It therefore came to no surprise too that the world, and for that matter Hong Kong, has not been ready to face any public health outbreak with efficiency.

Yet there had been warning signs. Plague returned to India in August 1994, and resistant T.B. became an epidemic in Russia in 2000. In 1998, WHO launched the Roll Back Malaria campaign to fund incentive for development of the anti-malaria drugs while chloroquine began to lose its effectiveness. Of course, there is the Avian Flu of 1997.

We have not taken the hint, we had lived in the comfortable ignorance oblivious to the fact that new human pathogenesis can emerge and old infectious once thought conquered could resurface with a vengeance.

In short, in the history of mankind, where there are victims, there will be infectious diseases. It becomes obvious therefore that we can look at infectious disease from three angles:

1. As a public health issue where properly organised public health means could prevent or control an epidemic, and where its failure could produce a catastrophy.
2. As a social issue, where the disease is known, the causative agent is identified, its prevention is well mapped, yet difficulties abound in convincing the society to adapt to it – from the Government to the man on the street, and even the world as a community to work together.
3. As a melodrama where the injection of politics, the struggle for power, could have blurred proper scientific investigations of the disease and hamper the precious lesions learned from any infectious outbreak that is to no ones' benefit. As Milton in his *Paradise Lost* said, "And out of good still to find means of evil".

Let me elaborate.

Whilst it may sound disheartening and perhaps even pitiful, it was decrease in infectious diseases which brought about the increase in life expectancy of the world not the discovery of curative medicine.

To wit, data from England, Wales and Sweden have shown that in 1700, the average male lived just 27 to 30 years. By 1971, male life expectancy was 75 years. More than half of that improvement occurred before 1900. In all 86 percent of the increased life expectancy was due to decrease in infectious diseases that occurred prior to the age of antibiotics. In the UK, T.B. death dropped from nearly 4,000 per million people to 500 per million between 1838 and 1949 (The year when antibiotics treatment was introduced). Since then, with the advent of anti-T.B. treatment in the next 20 years, the death rate only fell to 460 per million.

What prompt the decrease in infectious disease is of course a matter of considerable academic debate. Yet one cannot ignore nor neglect that the following could well be key issues: nutrition, housing, sewage disposal, safe drinking water, epidemics control, swamp drainage, public education, literacy, access to prenatal and maternity care. All and all public health issues.

A study of the story of plague will give the whole issue away. As known to us now, plague is caused by Yersinia pestis, a gram negative bacillus that live on fleas that parasite on black rats "*Ratus ratus*". Transmission to human is either through flea bite producing bubonic plague or in the case of pneumonic plague, from droplets – bacilli coughing out with blood from those infected. We know now too that the bacilli response favourably to tetracycline.

Plague in Hong Kong began in the early summer of 1894 before the antibiotic era. Many speculations surrounded the cause of the disease – from supernatural believe of offending the ancestors to obnoxious gas from the earth. By July of the same year the causative organism was discovered by Kitasato and Yersin, yet what could be done? Hygiene or better public health was perhaps the only action which ultimately found to be effective. The infection was highest in areas of Hong Kong where sanitation appeared to be the worst. The Sanitation Board ordered cleaning the streets in Taipingshan area, house to house search for sick and suspected patients, isolating these victims in three hospitals – Kennedy Town Police Hospital, the Glass Work Hospital controlled by the Tung Wah Board and a naval ship in the harbour – the Hygia. This was not without opposition and antagonism of the local population. People took to the streets – not uncommon by today's standard – chanting unfounded rumours that house search was an excuse for rape and pillage, and immediate removal of body for burial is for Westerner to remove body organs to grind up for medicine. The actions, though unpleasant, took effect and the epidemic was finally controlled by August 22. All in all, there were 2,679 cases notified, 2,552 died. Public health took centre page.

Nor was the story different in recent days – In the summer of 1994 following an earthquake in Maharashastra of India, plague broke out in a near town of Surat, a place suddenly overpopulated by migrant workers for diamond industry where sanitation was the exception rather than the rules.

Modern medical treatment did little to help to curb the epidemic. In fact, there was a general rush for tetracycline which were soon depleted and the Indian FDA was compelled to warehouse caches of the medicine.

What brought the containment was the declaration of Surat being "plague list" by the then prime minister – where army was dispatched to maintain order and quarantine, to stop exodus to other parts of the country, to burn up all waste, improve sanitation, kill all the rats. Again, a public health triumph.

Like the public medical service in Hong Kong during SARS, the Indian Public Medical Services all kept their ground and worked closely with private physicians. But unlike the dedication in Hong Kong, 80% of the private physicians in India went into panic and fled the city.

But public health is not the be-all and end-all for infectious diseases. In some of the worst infectious disease pandemic, social and societal issue takes the front page. Such as the case with HIV/AIDS.

Since the first case of HIV/AIDS was identified in 1981, and since the discovery of the retrovirus, most countries were aware of the mode of spread – through unprotected sex, sharing needles, mother to foetus transmission, accidental transfusion of infected blood. Similarly, most in the developed world would know the best way to prevent getting infected. The discovery of the "cocktail" treatment using protease inhibitors also brought new hopes to the HIV positive in that the treatment protocol taken lifelong could delay the onset of eruption of the disease – HIV positive but not AIDS affected.

Yet up to now some 20 years, the number of HIV/AID around the world was in no way curb. Instead, the trend is increasing, in particular in Sub-Saharan Africa, in the Indian subcontinent and in South East Asia – in China. Why did all these happen when the message is not about being discrete, but about using condoms and not sharing needles and syringes? Today, as the figure looms at 45 million, AIDS becomes not just a public health issue, not just a medical issue but a societal and community issue. There are at least five contributing factors.

Firstly, there is the issue of denial. In the Mainland, for example, HIV/AIDS was not taken on board as a national problem until as late as the early 1990's. Hitherto, HIV/AIDS was a foreigners' disease. Much change in attitude has since taken place for the better, hopefully it is never too late.

Secondly, there is the issue of stigmatization and discrimination. For whilst most people are conscious of the fact that HIV/AIDS cannot and will not be caught under the usually social style of contact, many still frowns on having an HIV/AIDS sharing the same building, walking the same road. It is a matter of proper education on a wide ranging base which could be extremely difficult considering the size of some countries and the remoteness of some isolated villages.

Thirdly, there is the discrimination between the rich and the poor. Yes, protease inhibitors are now available to delay the development of full blown AIDS, yet it has a price tag of over HK$1,000 a month. It became obvious that the treatment is for the very rich.

The state of affairs in the Sub-Saharan African Continent is a shining example. In South Africa for example, where the total infection rate is around five million, less than a few percentage can afford the treatment. Data shows that some 290 million Africans have an average income of less than US$1 per day. Whilst it is well known that the "cocktail treatment", if given at early pregnancy, could decrease the incident if not eliminate the chance of mother to foetal spread, such treatment was denied in favour of cost. Whilst it may be understandable that many developing or less developed countries could not afford to provide free "cocktail treatment" for all HIV positive victims, pharmaceutical industries are NOT willing to cut the cost, under the disguise of the need to recover the cost of research and development. It comes as no surprise that whilst the whole world in the issue of AIDS is ONE WORLD, with ONE HOPE, in many countries, it is obvious that we are all in one world, but many have no hopes. Ironically, it is in the poor countries that AIDS are most extensive.

The fourth issue is lack of trust, lack of trust in the personnel who are supposed to look after them, perhaps through lack of communication skills and ultimately reflecting the lack of confidence of the Government.

The tragedy in Henan (河南) is a vivid example of poverty, ignorance, lack of trust and discrimination. The Shangcai County (上蔡縣) in Henan is a very poor village. Farmers reap barely enough from their crops for daily living. Extra spending, even schooling for children, requires other ways of acquiring cash. Selling blood became the natural and easy source of income as blood is a "tool with no cost" (無本生利的工具). Much of the collected blood were pooled together and after the plasma proteins and other necessary blood elements were removed for other purposes, the left behind was transfused back to the donors – sparking a chain of infection transfer. Yes, through international agencies and the Central Government, drugs are available for most inflicted. Regrettably, the issuing of these "good wills" were never properly explained. Like all medications, some may experience certain side effects albeit minor, such as vomiting at the initial phase. Cynicism prevailed, many refuse to take the medication. A philanthropic act had thus turned sour. Worse, many look to hiding for fear of discrimination.

The fifth issue is that of societal priorities. As mentioned early, the initial phase of AIDS in the Mainland was that of denial – it is a foreigner disease – it was so sad. The leadership was never seen then to support the necessity to raise concern to AIDS. Yes, in January 1997, the late Minister of Health, Chen Minzhang (陳敏章) and myself visited the Ditan Hospital (地壇醫院) and shook hands with a HIV/AIDS patient. There were minimal and only local publicity. The visit of Bill Clinton speaking to Tsing Hua University (清華大學) on AIDS and afterwards shook the hands of an HIV patient and advocate raised the profile. Premier Wen(溫家寶) saw the need to visit AIDS inmates and place Wu Yi (吳儀) to take charge of AIDS policy and movement in China. For AIDS, both workers and victims, such was a triumph in politics.

But not all political move, albeit calculated, could end up in triumph as the story of SARS and its aftermath unfolds.

SARS, like a whirlwind, swept Hong Kong off its feet. Everything came almost to a standstill. The health care profession and health care services were hard hit most. Incidentally, we were completely ignorant of the cause, the way the illness spread, how we could protect ourselves and our patients, nor did not know the right form of treatment. It was a fear and frustration seeing a continuous stream of patients being admitted, one by one your working partners fall victims of the disease, not knowing when it is your own turn. But our health care workers braved on. Within weeks the causative Corona Virus was discovered and isolated. Every health care personnel stood firm, there was not a single deserter. Instead, some even volunteered to serve the infected wards to substitute or replace their colleagues who were either sick or because of higher risk. We lived with our masks, assumed a faceless status, gave up social life but with one aim in mind – get rid of that despicable infectious disease – regrettable, even as of today, there was no vaccine and no recognizable recommended treatment. It was a story of pure and profound bravery.

As SARS rages on, and as more data and statistics accumulated, Hong Kong's health care workers took the centrefold. Our detailed and transparent records, our rapid breakthrough in discovering the virus and the detail scientific studies on this new atypical pneumonia which could well be another pandemic, became the envy of the world. Our system for contact tracing, our direction to carry on long term studies of the patients for possible complications either of the disease itself or the treatment has put Hong Kong on the map to lead studies on infectious diseases. So far, over 50 papers have been published on SARS from Hong Kong for us to take the grid position.

Like other epidemics, SARS have brought Hong Kong back to realize that infectious diseases are still very important and that we have been inadequately prepared. It gave us the direction to rebuild the health care service and system, and to define our priorities in service and in training personnel. It also gave Hong Kong public medical services a chance to wrangle more money from the Government for health care to top up the needs for those priorities, and we have succeeded. A total of HK$1,165 million has so far been pumped into or approved for the public medical service to enrich health care in infection control and more could be forthcoming.

The final chapter of SARS could therefore be that of "they live happily ever after" or like the typical Chinese movie: 大團圓大結局.

But this is not to be the case, instead the closing chapters of the "activities that the whole world praised" were that of tears, of frustration, of rolling of heads and of unnecessary financial loss.

Tears, were uncontrollable to mourn the 299 who lost their lives, some eight of them from the health care team. Frustration, were from health care workers who, in spite of selfless devotion fighting the unknown, were critically interrogated by the Legislative Council Select Committee and criticized by its reports.

At least three in the hierarchy of health care resigned and one left Hong Kong seeking greener pastures. Millions of dollars were spent, perhaps unnecessary by the Legislative Council to stage the inquisition and by the Hospital Authority to prepare the data and legal grounds for the response. All these as one would realize are money from the public. But this is not all, legal battles are expected from the victims of SARS to the public health care service and possible compensation claims could be magnanimous.

Why did all this happen? Dare I say, it is all backfired from presumably calculated political moves.

The Government was criticized for initially playing down the severity of the infectious spread suggesting that there might be some "covering up" in favour of possible effects on Hong Kong's economy. But then what could the relevant Bureau do when even on April 22, 2003 the then China's Minister of Health at a press briefing in this very building to the Hong Kong media, said that there were only "a few cases of atypical pneumonia in China"?

It is obvious that a saga such as SARS where some 299 died and some 1,755 were infected and where Hong Kong was almost brought to a standstill deserved some form of a high power independent investigation, especially in this political climate of "blame culture" and "accountability and responsibility". There were calls therefore, from many, that an Independent Judicial Commission be set up by the Hong Kong Government to look at the whole picture – from health care to close of schools, quarantine to border control, and to suggest recommendations for the future.

This was regrettably not to be, in the interim, an expert committee was set up within the Health, Welfare and Food Bureau and an independent internal investigatory committee within the Hospital Authority, both to look at lessons to be learnt.

It is evident from the start that these two bodies and their expected reports other than an independent commission will not satisfy our "Peoples' Representatives" – paving the way for them to set up a Select Committee with "power and privilege" to call and "question witness". The aim of this Select Committee, since the Legislative Council was not satisfied with the other two reports that no blame was apportioned, was to "witch hunt for person or persons responsible".

Regrettably, this "Select Committee" had areas of fallacies.

As a start, the efficiency of this body stands to question. It is expected to investigate a complete unknown saga from medical causes and yet it did not have expertise of that specialty within its membership or even as advisors.

Secondly, much questioning were done on the frontline health workers with a motivation of criticism when it is obvious that whilst these frontlines were sacrificing their lives to fight the war of SARS, many of the members of the Select Committee were in the save environment of their offices, oblivious to the danger that their voters and constituents were facing. It was no surprise that at least one frontline executive wrote to the Select Committee, "Where were you when we needed you most".

Thirdly, perhaps in an attempt to lobby for votes for the then forthcoming Legislative Council election, conclusions of the Select Committee were leaked well before the investigation was completed, not just once but at least twice. An investigation was staged in the Legislative Council for the so-called "McDonald's" incidences, but as expected, produced no results. The second leak was from an obvious source but was side stepped by putting the blame on the "efficiency" of the Hong Kong Post Office.

What then was the subsequent score?

As mentioned earlier, three from the health care hierarchy have resigned and one left Hong Kong. Of course, they could be replaced.

Of the 11 members of the Select Committee, only four were returned to the Legislative Council, of the remaining seven, two elected to step down and five were defected. They were all replaced.

In the whole exercise, nobody won, both the hunters and the hunted succumb. In many cases, the hunter becomes the hunted.

The final closing chapter of SARS was thus a modern day *Hamlet*. Is it a comedy or a tragedy? Nay, it is neither, it is a pure melodrama of miscalculated politics.

But is the profession more united? Has the image of the profession improved? Is the integrity of the health care institutions maintained?

Alec McFadzean, I am confident, would NOT have approved.

Thank you!

Appendix 2

Useful web links

Fit for Purpose: A review of governance and management structures at The University of Hong Kong

https://www.hku.hk/about/governance/purpose_report.html

The Report of The Review Panel on the Centenary Ceremony held on August 18, 2011

http://www.gs.hku.hk/rpanel/Report.pdf

Standard Working Hours Committee Report

www.labour.gov.hk/eng/plan/pdf/whp/swhc_report.pdf